TECHNICAL
REPORT

Reducing Attrition in Selected Air Force Training Pipelines

Thomas Manacapilli, Carl F. Matthies, Louis W. Miller,
Paul Howe, P. J. Perez, Chaitra M. Hardison,
Hugh G. Massey, Jerald Greenberg,
Christopher Beighley, Carra S. Sims

Prepared for the United States Air Force

RAND PROJECT AIR FORCE

The research described in this report was sponsored by the United States Air Force under Contract FA7014-06-C-0001. Further information may be obtained from the Strategic Planning Division, Directorate of Plans, Hq USAF.

Library of Congress Cataloging-in-Publication Data

Reducing attrition in selected Air Force training pipelines / Thomas Manacapilli ... [et al.].
p. cm.
Includes bibliographical references.
ISBN 978-0-8330-5261-2 (pbk. : alk. paper)
1. Airmen—United States—Training of. 2. United States. Air Force—Officers—Training of. 3. United States. Air Force. Air Education and Training Command. 4. Technical education—United States. 5. Military education—United States. I. Manacapilli, Thomas.

UG638.R43 2012
358.4'150973—dc23

2011050774

Published 2012 by the RAND Corporation
1776 Main Street, P.O. Box 2138, Santa Monica, CA 90407-2138
1200 South Hayes Street, Arlington, VA 22202-5050
4570 Fifth Avenue, Suite 600, Pittsburgh, PA 15213-2665
RAND URL: http://www.rand.org/
To order RAND documents or to obtain additional information, contact
Distribution Services: Telephone: (310) 451-7002;
Fax: (310) 451-6915; Email: order@rand.org

Preface

The U.S. Air Force has a continuing interest in reducing high rates of attrition and training-block failures (known as *washbacks*, in which portions of training must be repeated) during initial skills training. High attrition requires a significantly greater input into the schoolhouses, increasing training and recruiting costs. High washback rates reduce the capacity of the schoolhouses by requiring that seats be set aside for students who need to retake training blocks, and such rates incur greater costs by increasing training time. The U.S. Air Force (USAF) Directorate of Force Management Policy (AF/A1P) tasked the RAND Corporation to investigate enlisted USAF specialties with high attrition and washback rates.

The research described in this report extended over two years, divided into two phases. The first phase was the investigation of five career fields defined by the following Air Force specialty codes (AFSCs): 1C1X1, air traffic control (ATC); 1C2X1, combat control (CCT); 1T2X1, pararescue (PJ); 2A6X2, aerospace ground equipment (AGE); and 3E8X1, explosive ordnance disposal (EOD). This research was conducted as part of a fiscal year (FY) 2008 umbrella study called "Enhanced Testing and Screening for High Value and High Attrition Programs." The second phase continued the investigation into four other career fields: AFSCs 1N0X1, operations intelligence; 1N4X1, network intelligence analysis; 1N3X4X, Far East linguist; and 1N3X5X, Middle East linguist. This research was conducted as part of an FY 2009 umbrella study, "Personnel Selection and Screening." Over the course of the research, RAND researchers briefed numerous senior decisionmakers on its findings. Consequently, many of the recommendations have been or are in the process of being implemented. The report credits the Air Force for changes made through 2010 but does not include changes made after 2010.

This report should be of interest to USAF leaders and staffs concerned with reducing attrition in initial skills training.

The research reported here was sponsored by AF/A1P with support from the Air Education and Training Command Directorate of Operations (AETC/A2/A3) and conducted within the Manpower, Personnel, and Training Program of RAND Project AIR FORCE (PAF).

RAND Project AIR FORCE

RAND Project AIR FORCE (PAF), a division of the RAND Corporation, is the U.S. Air Force's federally funded research and development center for studies and analyses. PAF provides the Air Force with independent analyses of policy alternatives affecting the development, employment, combat readiness, and support of current and future air, space, and cyber forces.

Research is conducted in four programs: Force Modernization and Employment; Manpower, Personnel, and Training; Resource Management; and Strategy and Doctrine.

Additional information about PAF is available on our Web site:
http://www.rand.org/paf/

Contents

Figures

Tables

Summary

Introduction

The initial skills training process in the U.S. Air Force, commonly called *schoolhouse* or *technical training*, is illustrated in Figure S.1 as part of the larger training pipeline. It begins when an individual is recruited. For recruits entering a majority of career fields (including those studied in this report), the career field is guaranteed during the recruiting process. The individual then enters the Air Force approximately nine weeks prior to seat availability in the training program. If an immediate seat is not available at the time an individual is recruited, the recruit is placed in the delayed-enlistment program. When a seat becomes available, the recruit ships to basic military training (BMT) for a little more than eight weeks of training. After completing BMT, the airman transfers to technical training for specialty-specific training. After graduating, he or she is awarded the status of apprentice, designated by a 3 skill level in his or her AFSC, and ships to his or her duty assignment for employment in the specialty and continued on-the-job training.

The focus here is on initial skills training but does not exclude the other areas where applicable. The Air Force has a continuing interest in reducing high rates of attrition and training-block failures (called *washbacks*)—in which portions of training must be repeated during initial skills training. High attrition requires significantly more student entries into the schoolhouses, increasing training and recruiting costs. High washback rates increase needed schoolhouse capacity by requiring that seats be set aside for students who need to retake training blocks, and the higher rates incur greater costs by increasing training time.

That concern is shown in Table S.1; in FY 2001–FY 2008, initial training costs amounted to about $1.3 billion per year. Attrition and washback rates averaged 8 percent and 21 percent, respectively, which led to attrition and washback costs of $112 million per year. Thus, reductions in attrition and washbacks in career fields with high rates can result in significant cost savings.

Figure S.1
Generic Training Pipeline

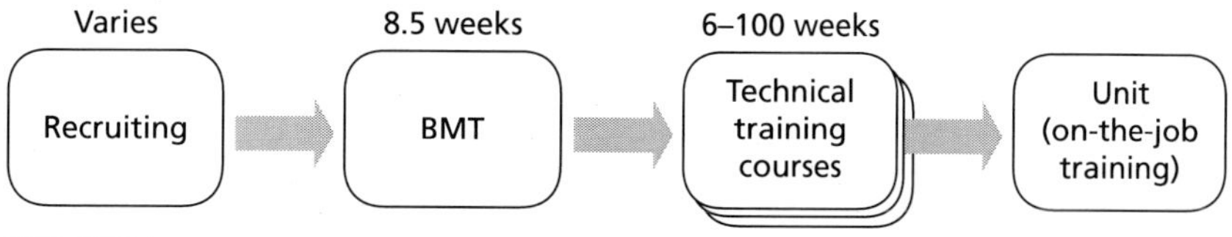

RAND TR933-S.1

Table S.1
Key Initial Training Factors

Category	Item	Factor
Initial training costs	Approximate annual number of recruits	34,000
	Average length of initial training	22 weeks
	Average training cost per recruit per week	$1,700
	Approximate cost of initial training	$1.3 billion[a]
Attrition costs[b]	Average attrition	8%
	Estimated cost of attrition	$67 million
Washback costs[c]	Average washback	21%
	Estimated cost of washback	$67 million

NOTE: Population and average lengths are based on FY 2008 data. Attrition and washback averages are based on FY 2001–FY 2008 data.

[a] The approximate cost of initial training is calculated by multiplying the approximate number of recruits, the average length of initial training in weeks, and the average training cost per recruit per week. Considering significant digits, 34,000 × 22 × $1,700 = $1.3 billion.

[b] Attrition costs include training individuals (half-length) who then attrit ($51 million) and the cost of recruiting additional personnel ($16 million).

[c] Washback costs are based on one-quarter of average training time.

Given these costs, AF/A1P asked PAF to examine and recommend methods for reducing training attrition in several enlisted career fields. AF/A1P's goal was to maximize the fiscal impact of implementing any recommendations we made, by selecting specialties with the highest attrition and washback costs.[1] Five initial skills training specialties were selected for examination: PJ, EOD, CCT, ATC, and AGE. In the following year, AF/A1P added four other specialties: operations intelligence, network intelligence analysis, Far East linguist, and Middle East linguist.

In approaching the task, we reviewed the extant literature, incorporating ideas as appropriate, and focused on two approaches. First, as an exploratory investigation into possible causes of attrition, we visited each of the training bases and talked with students and instructors about reasons for high attrition. Second, in a parallel effort, we evaluated alternative selection tools that might reduce attrition. We combined multiple USAF databases to determine relationships between personnel data and training outcomes in the subject AFSCs. We used regression techniques to develop mathematical models, which evaluated the usefulness of these tools. Additionally, the interviews led to other analytic approaches to investigate attrition causes or alternative selection tools.

We also reviewed corollary components of the accession pipeline (BMT, recruiting, and classifying) and talked with subject-matter experts in each of these areas.

---

[1] Attrition occurs when the student fails to complete the course. Attrition can occur at any point in the training. Attrition reasons can vary from academic failure to health issues to behavior concerns. Washbacks occur when the student has to repeat some portion (also called a *block*) of the course. Some individuals wash back numerous times before graduating or attriting.

Over the course of the two-year research period, we briefed numerous senior decision-makers on our findings. Consequently, many of the recommendations have been or are in the process of being implemented.

The results of the analysis fall into two categories: specialty-specific findings and cross-cutting findings.

Specialty-Specific Findings

During the analysis, we discovered that each specialty we examined had issues specific to its training.

Air Traffic Control, AFSC 1C1X1

Over the years, much research has been accomplished recommending the use of a screening test for ATC selection.[2] A test used by the Federal Aviation Administration (FAA) would require an hour and a half to administer and might be accomplished at the recruiter's office rather than the Military Entrance Processing Station (MEPS).[3] This would not be unusual because the Air Force conducts the physical ability and stamina test (PAST) at individual recruiter locations.

Combat Control, AFSC 1C2X1

Self-initiated elimination (SIE), the USAF term for voluntary disenrollment, accounted for 75 percent of all attrition from the CCT pipeline at Lackland and Keesler Air Force Bases (AFBs) between October 2006 and April 2008. The majority of SIE cases were rooted in the student's belief that CCT was not an appropriate specialty for him.[4] This suggests that more information is needed at the recruiter's office that explains what the job entails.

There appears to be some correlation between officers training with the team and a lower attrition rate. The data obtained from the schoolhouse were sparse but showed a statistically significant 20-percent increase in retention. We suggest a small test of spacing officer trainees out one to a class.

Another cause of CCT attrition concerns physical fitness. We discuss these observations about physical fitness and the Fit Flight concept under PJ findings later in this summary.

Operations Intelligence, AFSC 1N0X1

Interviews with students revealed that some instructors routinely leave the students in unsupervised—and reportedly unproductive—study periods during a majority of the blocks. This course is a candidate for self-paced learning because some students can easily memorize the information and finish the program much more quickly than others can. Additionally, the lack of supervised study time results in a poor atmosphere for some students trying to study and memorize information.

---

[2] We identified a dozen such studies.

[3] Classification testing, in the form of the Armed Services Vocational Aptitude Battery (ASVAB), is completed at a MEPS, but the test would have to be approved by the Manpower and Accession Policy Working Group. A 1.5-hour test would not likely be approved. Therefore, the Air Force is testing a shortened version. Additionally, the test would have to be administered by proctors to maintain test security.

[4] PJ and CCT specialties are open to men only.

During the period of the research, a few instructors were first-assignment instructors, having gone straight from training to instructor. The students, especially prior-service (PS) students, found these instructors to be subpar, lacking experience to back up the instruction.

Interviews of instructors and students suggested that the specialty description is not representative and might glamorize the job. They also confirmed that some academic failures are deliberate schemes to get reclassified.

The training requires a lot of memorization, and the job requires public-speaking ability, suggesting that this specialty might be amenable to a screening test, possibly an oral one.

Far East and Middle East Linguists, AFSCs 1N3X4X and 1N3X5X

Interviews with students and instructors resoundingly emphasized the importance of motivation. Dornyei (2001) confirmed the importance of enthusiasm, commitment, and perseverance in learning a foreign language. The length and intensity of the phase program decrease motivation.[5] Students did not feel that the military training leaders (MTLs), with no experience learning a language, understood the difficulty of the program.[6]

Language requirements are determined by operational needs. And not getting a preferred language can also decrease an individual's motivation. Yet, training squadrons did not have the authority to allow students with similar Defense Language Aptitude Battery (DLAB) scores to make mutually agreeable language switches.

The Defense Language Proficiency Test (DLPT) is an all-or-nothing test at the end of the yearlong program. Squadron statistics showed that, over the past three fiscal years, 76 percent of airmen completed the basic course of study and that approximately two-thirds of these students passed the DLPT at proficient levels, meaning that the overall completion rate was near 50 percent—higher in some languages, lower in others. This pass rate, among the Air Force's most highly qualified airmen (as measured by the ASVAB), leads us to question whether expectations are too high. The DLPT is under recalibration, but, if the most-qualified recruits are failing at a 50-percent rate, are expectations too high? Is this specialty too demanding for enlisted personnel? Or could students who fail to reach required language proficiency be used in some restricted mode and continue to learn the language while on the job?

Network Intelligence Analysis, AFSC 1N4X1

The primary concern of students was the length of the training: approximately five months. Many students felt that the pace was slow, despite the tremendous amount of information to absorb, and that some material would be better deferred to on-the-job training. Self-pacing might be an option in this course. Also, studying is difficult because of the classified nature of the work and the need to go to a secure facility to study classified material.

Additionally, at the time they were recruited and classified, many students did not understand what the specialty involved. Understandably, many aspects of the specialty are classified, but students felt that a better description of the specialty was possible.

---

5 The phase program is designed to gradually take recruits from the very controlled environment they recently experienced in BMT to a less strict environment via a series of graduated privileges granted over an extended period. The privileges do offer relief from the BMT environment but never equal the privileges that an airman has in the operational Air Force. Consequently, over time, they have a negative impact on morale.

6 MTLs supervise and execute the military phase of the instruction (nonnonvocational training), which is a continuation of the training received in BMT.

Aerospace Ground Equipment, AFSC 2A6X2

AGE attrition, 4.4 percent in FY 2007, has dropped from double-digit levels (FY 2002–FY 2004) to single-digit levels in the past few years.[7] Washbacks remain high, ranging from 17.1 percent to 30.3 percent.[8] We found no direct correlation between washbacks and attrition, although one might infer a relationship. We calculated that, for this AFSC, one attrited individual costs the same as 2.2 washbacks, and, if AGE cut its washback rate in half (down to 15 percent) by eliminating rather than washing back personnel, the attrition rate could increase to 10.9 percent without any increase in costs.

Training squadron (TRS) personnel worked hard to reduce previously high attrition levels. They reviewed the reading level of materials, instructor performance, and test questions and took specific steps to fix problems identified internally.

Our interviews revealed that the course might be longer than necessary. This was the majority view among the instructors interviewed. Without any prompting, students also volunteered opinions that the course was too long. There is some correlation between course length and attrition. We would not expect much savings in attrition costs because of a shorter course, but we might see savings in reduced washbacks and would definitely see overall savings in basic course costs from eliminated training days and associated expenses.

Pararescue, AFSC 1T2X1

The PJ career field eliminates more than 70 percent of the entries into its multicourse pipeline. Fortunately, most of the eliminations occur in the indoctrination course, the very first course in the pipeline. If attrition is going to be high, and it historically has been, then this is the most cost-effective place for attrition to occur.

Previous research shows that higher levels of physical fitness are correlated with success in the indoctrination course. The majority of students interviewed felt that their fitness levels declined during BMT. Some mentioned being ridiculed by their BMT instructors for attempting extra push-ups. The Air Force has instituted Fit Flights, designated by blue armbands, for certain career fields with exceptional physical demands. However, the members of Fit Flights are interspersed among regular trainees and do not engage in additional physical exercise. We believe that the concept of Fit Flights is good but that Fit Flights should be segregated for physical training (PT) to conduct PT at a higher level. Although this might engender elitist attitudes, we found no evidence that this would be bad for battlefield-airman training.[9]

Explosive Ordnance Disposal, AFSC 3E8X1

EOD attrition has averaged 33 percent over the past eight years in the apprentice course at Eglin AFB. The primary reason for failure is academic deficiency (77 percent); the second-largest reason is voluntary disenrollment (14 percent). The Air Force has the highest attrition rate of all four services at this joint school. This appears directly related to the proportion of non-PS (NPS) trainees: More than three-quarters of USAF trainees are NPS, while sister services are more likely to send midcareer personnel. When compared by pay grade, the USAF

---

[7] Latest attrition figures for AGE were 4.1 percent for FY 2008 and, as of July 1, 2009, 5.6 percent for FY 2009.

[8] Washbacks for AGE were 15.5 percent in FY 2008 and, as of July 1, 2009, were 23.4 percent in FY 2009.

[9] Battlefield airmen are also referred to as *special forces* in the generic military use of the term.

attrition rate is similar to that of the Army. Increasing the ratio of PS students should lower the overall attrition rate.

The Air Force also utilizes the shortest preparation program prior to EOD training. While the Army and Navy focus on preparation (a ten-week course), the Air Force uses a preliminary course (six days), effectively a screening program, prior to sending its trainees to apprentice training. The Army program is longer than the USAF program but does not result in much difference in the overall attrition rate. Navy attrition over its preliminary program is similar to USAF attrition (both approximately 50 to 60 percent). But the Navy program reduces attrition considerably at the joint school. Although the Air Force could reduce attrition by increasing the preliminary-school length, greater savings can be realized by changing the proportion of PS students.

Previous research indicates that the use of noncognitive screening tools can reduce attrition. The Air Force is currently conducting research on one such tool, the Emotional Quotient Inventory (EQ-i®). Following our literature review, we recommended the use of a noncognitive test, potentially the EQ-i, for screening entrants into this specialty.

Cross-Cutting Findings

We found that some issues were applicable to the majority of specialties we examined.

Recruiting

In each of the nine specialties, 20 percent or more of the trainees specifically mentioned getting poor information from their recruiters. The majority found information from friends, family, and the Internet (YouTube was frequently mentioned). A majority of trainees agreed that more and better information from recruiters would have been helpful, especially in terms of what the training would be like, allowing them to mentally prepare themselves for the demands of the training. There were some cases of exemplary recruiter assistance, such as putting recruits in touch with individuals in their prospective career fields. We could not directly estimate the impact that better information from the recruiters had on retention, but, with 75 percent of CCTs voluntarily disenrolling, it could be substantial. The interviews included unsubstantiated instances of recruiters gaming the system by suggesting to recruits that they could sign up for the CCT specialty to get into the Air Force quickly and then self-eliminate if they did not like it.

Phase Program

The student interviews revealed grievances with the phase program, a schedule of restrictions designed to maintain discipline after BMT.[10] The longer an individual had been in the phase program, the greater the resentment he or she expressed during the interviews. We believe that the phase program is good overall and should not be discontinued. We believe that the majority of MTLs do not abuse their position of authority, as some were alleged to have done. But we also believe that changes to the phase program are warranted to prevent the few excesses that do occur. Those misuses of authority ultimately find their way home (literally) to negatively

[10] The phase program, as executed during the course of this research, consists of three separate phases, with decreasing restrictions as one progresses through the phases.

affect USAF recruitment. In addition, the system is naturally perceived as unfair. Students in shorter courses graduate from the phase program sooner and consequently enjoy a more lenient disciplinary environment at their first duty station, which, in effect, penalizes airmen who are enrolled in longer training courses (typically airmen with higher ASVAB scores). Finally, although the phase program probably does not directly cause attrition, it is a key factor in adding stress to the recruit's daily life. Ultimately, any additional stress, such as financial difficulties or family problems, can overwhelm the recruit, destroying any motivation to finish the program.

We recommend multiple changes to phase III. At a minimum, at some point, it should not be possible to send an individual back to phase I, barring an Article 15 or some serious infraction.[11] Additionally, we recommend making the phase program equal length for all specialties and suggest six to eight weeks as an appropriate length.

Training-Base Purpose

At Lackland, the base is designed around the recruit. Many service and support facilities are open outside the hours of the training day. At all other training bases, the base operates on hours like a nontraining base, with limited availability of services and support after the duty day. For routine medical appointments, other appointments, and errands requiring more time than the one-hour lunch break, the trainee must be excused from training. In many cases, this is a contributing factor in washbacks and, potentially, in attrition. Even simple tasks, such as sending a package through the post office during the lunch hour, can be complicated by the fact that all the other trainees are trying to do the same thing. We recommend that, to serve the trainee, training bases consider adjusting the work schedules of clinics, administrative services, postal services, and those performing clothing alterations.

Linked to training-base hours is the length of the training day. Some of the instructors interviewed suggested that information covered in eight hours could just as easily be taught in six. Reducing the training day to six hours would provide the trainees with time to run errands and make clinic appointments without missing instruction. There is also a question of trainees retaining material taught in an eight-hour day. Some squadrons use the extra two hours for hands-on and simulation activity to reinforce what was taught during the first part of the day. The extra two hours would also provide some needed down time to trainees, a comment heard repeatedly in the student interviews.

Documentation

It became apparent during the course of the research that Air Education and Training Command (AETC) Form 125A, which is used to record the reason for attrition, was not used consistently among bases.

Areas for Further Research

We did not research BMT, but we did talk with BMT personnel at the 319th Classification Squadron. We learned that medical attrition is high and that individuals in a holding status

[11] Article 15 of the Uniform Code of Military Justice defines the commanding officer's options for nonjudicial punishment. See U.S. Department of Defense (DoD), undated.

attrit at a high rate. Part of the reason is that individuals in hold for discipline issues are grouped together with medical holds, and the medical holds receive little motivation to continue in the Air Force. We heard that the Army has substantially reduced its medical-hold attrition, but we did not investigate those claims. We recommend further research into medical holds.

We also reviewed the program used to classify individuals. Our discussion with users indicated that the model is sensitive to weights assigned to various factors. Although we recognized that it is impossible to satisfy every person's specialty preference, there might be room for improvement. We recommend further research into specialty classification.

Recommendations

Specialty-Specific Recommendations

- Shortening the AGE course should be considered.
- Entries to the EOD specialty should include a higher percentage of PS trainees; a noncognitive screener should be used.
- The Fit Flight concept for PJ and CCT recruits should be continued. Fit Flights should be populated entirely by Fit Flight recruits with additional physical training. A few battlefield airmen should be used as training instructors in BMT.
- Officers attending combat rescue officer training should be spaced out such that at least one officer is in each CCT training flight at Keesler.
- An ATC screener should be used for classification in both ATC and CCT specialties.
- Operations intelligence and network intelligence courses should be reviewed for possible shortening. First-assignment instructors should not be utilized for training. Self-pacing for memorization-heavy training should be used.
- Linguist training placement should be improved to match more airmen into preferred languages.
- Further research is needed to determine whether airmen who score at less-than-proficient levels after completing language training can reach proficiency through on-the-job training.

General Recommendations

- Modify the phase program, potentially adding a phase IIIb at six weeks from which recruits would not be phased back unless a serious incident occurs (Article 15 nature). Another alternative would be to end the phase program for all AFSCs at six to eight weeks.
- Have training bases adjust the service and support facility operating hours around the trainee, or reinstate the six-hour training day.
- Have career field managers (CFMs) provide more information to recruiters about the specialty, the training, and how to be prepared.[12]
- Conduct further research into medical holds at BMT and the classification process.
- Guidance for Form 125A packages should be clarified because inconsistencies among the bases made them difficult to analyze.

[12] The CFM provides central oversight for career-field education and training, chairs the utilization and training workshops, and addresses day-to-day career-field operations issues and specialty concerns.

Acknowledgments

We have received tremendous support from AETC, 2nd Air Force (2AF), Air Force Recruiting Service (AFRS), and all the groups and squadrons we visited. We thank the following from headquarters AETC: Maj Gen Mark Zamzow, Maj Gen Richard Devereaux, Col Scott Bethel, Lt Col Scott Long, Andrew Donate, Oscar Vigil, Douglas Mettler, Mark Pennington, and Gina Johnson. Colonel Bethel especially helped in opening doors for us to visit units within AETC.

At 2AF, Maj Gen Alfred Flowers, Maj Gen Michael Gould, Col David Orr, Walter E. Hack Jr., and James Hollingsworth were extremely helpful and supportive. Colonel Orr provided a lot of support as well in setting up visits. At the end of the research, Maj Gen Mary Kay Hertog and her staff reviewed the report and provided helpful information on recent USAF actions to address attrition issues.

At the Occupational Measurement Squadron, Capt Bryan Calkin supported our effort with access to various databases.

At AFRS, Col Earl S. "Stan" Chase and Paul Ditullio worked with us to provide recruiting data.

Air Staff personnel John Park, Lisa Hughes, CMSgt Donald Cantwell, and SMSgt Kelly Downey all assisted with direction, information, and advice.

Other CFMs, CMSgt Robert Hodges and CMSgt Lawrence Cole, also supported us with contacts and information.

At Lackland AFB, Col Richard Naylor, Lt Col Anthony Capobianco, Edward Lundberg, William Grozdanich, Ray Gutierrez, Neal Baumgartner, TSgt Keith Marshall, and Arthur Kupferer were especially helpful. Dr. Baumgartner provided physical-fitness analyses. Mr. Kupferer accessed and copied hundreds of Form 125A packages. Colonel Capobianco provided full access to his squadron and detachments.

Also at Lackland AFB, Lt Col Daniel Edwards, Bob Wilson, Laura Munro, Capt Donna Radcliff, Margaret Skuravy, and SMSgt Tonya Harris provided insight and data on BMT.

At Sheppard AFB, Col Scott Kelly, Maj Paul Abair, Capt James Poteat, MSgt Richard Vorbeck, TSgt Patrick Gesick, SSgt Phillip Davey, Philip Noe, Patsy Klein, Michael Dow, Russell Mancusco, William Norton, Terry Owens, Judith Sullivan, Marlene Fanning, and Risa Hillard were all very helpful and professional. Major Abair and Sergeant Vorbeck were especially helpful in locating and providing supporting data.

At Keesler AFB, Lt Col Scott Chowning, Maj Jesse Murray, Capt Daniel Stimpfel, SMSgt CarolAnn Daniels, TSgt Jarmin Blanton, TSgt John Wylie, TSgt Michael White, TSgt Marsha Kurley, SSgt Phyllis Amos, Ronald Wines, Steve Harberson, Danny Cable, and

Michael Moroney greatly assisted the research with insights and support. Colonel Chowning, Sergeant Daniels, and Sergeant Blanton went above and beyond with their support.

At Eglin AFB, Maj Jason Warnick, MSgt Timothy Sprosty, TSgt Beniquez, MSgt Lockhart, TSgt Fleming, TSgt Frederickson, SSgt Styls, and SSgt Kelch provided many useful insights and support for our effort. Major Warnick gave us a tremendous amount of his time before, during, and after our visit. We are grateful to Maj Tom Bongiovi, with whom we never met or spoke, as his previous research on EOD training was extremely important to the research.

At Kirtland AFB, Capt Mark Van Weezendonk and his staff were a great help in gathering personnel for interviews and briefing us on the pararescue training pipeline.

At Goodfellow AFB, Col Anthony Lombardo, LtCol Thomas Hensley, Lt Col Thomas W. Dobbs, and Maj Sammy Vega offered insights on the unique aspects of intelligence training and provided a very full and fruitful three-day visit.

At the Defense Language Institute Foreign Language Center in Monterey, California, Lt Col Paul Issler and Lt Col Jon Rice provided full access to their training squadrons, while Maj Donald Brunk and Maj Brian Garino coordinated all aspects of our visit. We also greatly benefited from the insight of leaders and instructors in the Arabic, Chinese, and Korean programs.

At the Air Force Personnel Center, Maj Gen Anthony Przybyslawski, Kenneth Schwartz, and Johnny Weissmuller assisted our research greatly. Mr. Schwartz traveled with us on some of our trips and provided many background data from ongoing research.

Thomas Carretta, Air Force Research Laboratory, has a long history of research into attrition for the Air Force. His previous research and his assistance during this research greatly benefited our effort.

We are grateful for Bruce Gould, who also assisted in our data gathering.

We also wish to thank John Crown, who was instrumental early on in the data analysis.

Finally, Lt Gen (ret.) David McIlvoy's and RAND senior analyst Harry J. Thie's reviews of the report made a tremendous improvement.

Abbreviations

1A8X1	airborne cryptologic language analyst
1C1X1	air traffic control
1C2X1	combat control
1N0X1	operations intelligence
1N1XX	imagery analysis
1N3X4X	Far East linguist
1N3X5X	Middle East linguist
1N3XX	ground-based cryptologic language analyst
1N4X1	network intelligence analysis
1T2X1	pararescue
2A6X2	aerospace ground equipment
2AF	2nd Air Force
3E8X1	explosive ordnance disposal
AAE	active enlisted master personnel extract
ACC	Air Combat Command
AETC	Air Education and Training Command
AETC/A2/A3	Air Education and Training Command Directorate of Operations
AETC/A3	Air Education and Training Command Directorate of Intelligence, Operations, and Nuclear Integration
AETCI	Air Education and Training Command instruction
AETC SAS	Air Education and Training Command Studies and Analysis Squadron
AF/A1	Air Force Personnel Directorate

AF/A1P	Directorate of Force Management Policy
AFAA	Air Force Audit Agency
AFB	Air Force base
AFI	Air Force instruction
AFMAN	Air Force manual
AFOQT	Air Force Officer Qualifying Text
AFPC	Air Force Personnel Center
AFRS	Air Force Recruiting Service
AFSC	Air Force specialty code
AFSOC	Air Force Special Operations Command
AGE	aerospace ground equipment
ANOVA	analysis of variance
APAS	Advanced Personnel Acquisition System
ASVAB	Armed Services Vocational Aptitude Battery
ATA	academic training adviser
ATC	air traffic control
AT-SAT	Air Traffic Selection and Training
BMI	body mass index
BMT	basic military training
CCT	combat control
CFETP	Career Field Education and Training Plans
CFM	career field manager
DLAB	Defense Language Aptitude Battery
DLI	Defense Language Institute
DLIFLC	Defense Language Institute Foreign Language Center
DLPT	Defense Language Proficiency Test
DLPT5	Defense Language Proficiency Test Five
DMDC	Defense Manpower Data Center
DoD	U.S. Department of Defense
EAD	enter active duty

EMT	emergency medical technician
EOD	explosive ordnance disposal
EQ-i	Emotional Quotient Inventory
FAA	Federal Aviation Administration
FY	fiscal year
GTEP	guaranteed training enlistment program
H2F	hard to fill
HQ	headquarters
HRRD	Human Resource Research Databank
HumRRO	Human Resources Research Organization
ISC	initial skills course
ITRO	interservice training organization
LOA	lack of ability
LOE	lack of effort
MAGE	mechanical, administrative, general, and electrical
MEPS	Military Entrance Processing Station
MLI	military language instructor
MTL	military training leader
NPS	non–prior service
PACE	processing and classification of enlistees
PAF	RAND Project AIR FORCE
PAST	physical ability and stamina test
PCS	permanent change of station
PCSM	Pilot Candidate Selection Method
PJ	pararescue
PJM	person–job match
PROMIS	Procurement Management Information System
PS	prior servicc
PT	physical training
SDI+	self-description inventory

SERE	survival, evasion, resistance, and escape
SIE	self-initiated elimination
TACP	tactical air control party
TBAS	Test of Basic Aviation Skills
TI	training instructor
TRG	training roup
TRS	training squadron
TTMS	Technical Training Management System
USAF	U.S. Air Force
USAFSAM	U.S. Air Force School of Aerospace Medicine
USMEPCOM	U.S. Military Entrance Processing Command

CHAPTER ONE

Introduction

Background

U.S. Air Force (USAF) initial skills training, also called *technical training*, is a part of a larger training process illustrated in Figure 1.1. It begins when an individual is recruited. For recruits entering a majority of career fields (including those studied in this report), the career field is guaranteed during the recruiting process. The individual then enters the Air Force approximately nine weeks prior to seat availability in his or her training program. If an immediate seat is not available at the time an individual is recruited, the recruit is placed in the delayed-enlistment program. When a seat becomes available, the recruit ships to basic military training (BMT) for a little more than eight weeks of training. After completing BMT, the airman transfers to technical training for specialty-specific training. After graduating, he or she is awarded the status of apprentice, designated by a 3 skill level in his or her Air Force specialty code (AFSC), and ships to his or her duty assignment for employment and continued on-the-job training.

The Air Force has a continuing interest in reducing high rates of attrition and training block failures (called *washbacks*)—in which portions of training must be repeated during initial skills training. High attrition requires significantly greater numbers of student entries into the schoolhouses, increasing training and recruiting costs. High washback rates reduce the capacity of the schoolhouses by requiring that seats be set aside for students who need to retake training blocks, and they incur greater costs by increasing training time.

That concern is shown in Table 1.1; based on data from FY 2001–FY 2008, the table shows that initial training costs amount to about $1.3 billion per year. Attrition and washback rates are 8 percent and 21 percent, respectively, which lead to attrition and washback costs totaling $134 million per year. Thus, reducing attrition and washbacks in specialties with high rates of both can result in significant cost savings.

Figure 1.1
Generic Training Pipeline

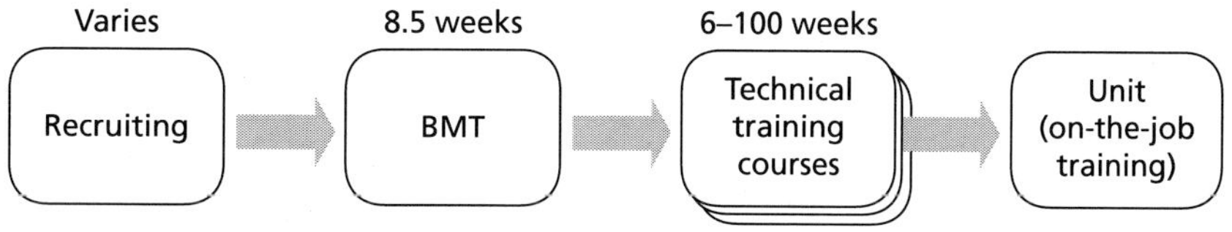

RAND *TR955-1.1*

Table 1.1
Key Initial Training Factors

Category	Item	Factor
Initial training costs	Approximate annual number of recruits	34,000
	Average length of initial training	22 weeks
	Average training cost per recruit per week	$1,700
	Approximate cost of initial training	$1.3 billion[a]
Attrition costs[b]	Average attrition	8%
	Estimated cost of attrition	$67 million
Washback costs[c]	Average washback	21%
	Estimated cost of washback	$67 million

NOTE: Population and average lengths are based on FY 2008 data. Attrition and washback averages are based on FY 2001–FY 2008 data.

[a] The approximate cost of initial training is calculated by multiplying the approximate number of recruits, the average length of initial training in weeks, and the average training cost per recruit per week. Considering significant digits, 34,000 × 22 × $1,700 = $1.3 billion.

[b] Attrition costs include training individuals (half-length) who then attrit ($51 million) and the cost of recruiting additional personnel ($16 million).

[c] Washback costs are based on one-quarter of average training time.

Objective

The USAF Directorate of Force Management Policy (AF/A1P) asked RAND Project AIR FORCE (PAF) to examine and recommend methods for reducing training attrition during initial skills training for several enlisted career fields.[1] AF/A1P's goal was to maximize the fiscal impact of the research by selecting specialties with the highest attrition and washback costs.[2] Initially, five USAF specialties were selected for examination in this research: pararescue (PJ), explosive ordnance disposal (EOD), combat control (CCT), air traffic control (ATC), and aerospace ground equipment (AGE). Historically, these specialties have had some of the highest attrition costs in the Air Force.[3] Later, four more specialties were added to the research: operations intelligence, network intelligence analysis, Far East linguist, and Middle East linguist.

[1] The focus of this research is initial skills training but does not exclude the other areas where applicable.

[2] Attrition occurs when the student fails to complete a course. Attrition can occur at any point in the training. Attrition reasons can vary from academic failure to health issues to behavior concerns. Washbacks occur when the student has to repeat some portion (also called a *block*) of a course. Some individuals wash back numerous times before graduating or attriting.

[3] The AGE career field has made great strides reducing attrition in recent years. AGE was chosen more for its high rate of training-block failures, known as *washbacks*.

Table 1.2
Actual Versus Programmed Attrition for Key Specialty Courses

Course Code	Course Title	Actual Attrition (%)	Programmed Attrition (%)
E3ABR1C131 00RB	Air Traffic Control Radar Apprentice	23.2	13
E3ABR1C131 00TB	Air Traffic Control Tower Apprentice	21.7	7
E3AQR1C231 00FC	Combat Control Operator Fundamentals	33.3	9
E3AQR1C231 00TB	Combat Control Operator Tower	29.2	14
L3ABP1C231 0C1A	Combat Control Apprentice	35.5	30
L3AQR1C231 0C0A	Combat Control Orientation Course	43.6	24
JBAQN3E831 00NA	Explosive Ordnance Disposal Apprentice	35.6	34
L3AQR3E831 0E0A	Explosive Ordnance Disposal Preliminary Course	29.2	47
L3AQR1T231 0P1A	Pararescue Indoctrination Course	60.9	70
L3AQR2A632 046B	Aerospace Ground Equipment	5.8	3
LCAQP1XXXX 0D0A	Air Force Combat Dive Course	36.4	4
X3ABR1N031 0A6A	Operations Intelligence Apprentice	21.4	—
XAABR1N431 0A1B	Network Intelligence Apprentice	12.5	—
XBAQZ1N334A0CMA	Apprentice Far East Crypto Logic Linguist	31.0	40
XBAQZ1N335L0PVA	Pashtu Basic	49.4	40

SOURCE: 2AF, as of October 2007.

NOTE: No information was available for programmed attrition for operations intelligence or network intelligence apprentice courses.

PAF was asked to identify possible causes of training attrition in each AFSC and to suggest remedies. Table 1.2 is a comparison of the actual versus programmed attrition rates for some of the courses that make up the training for these nine specialties.[4]

Training Process

Figure 1.1 shows the process of training for a generic specialty. There are other processes involved in transforming a recruit to a trained airman. The process of recruiting and selecting individuals differs for each of the nine specialties chosen here. In eight of the nine specialties studied, most non–prior-service (NPS) entries are through the guaranteed training enlistment program (GTEP), meaning that the selection of the person for the job occurred during recruiting and is included in the enlistment contract between the Air Force and the individual. In the PJ and CCT specialties (both 100-percent GTEP), a recruiter administers the physical

[4] Programmed rates are based on a two-year historical average of actual rates and are used for budgeting purposes.

ability and stamina test (PAST) to determine the individual's physical suitability for the job.[5] This test consists of running, swimming, and calisthenic events and is a key determinant in a person's ability to complete training in the most-strenuous AFSCs. The next stage in the process is BMT, during which attitudes toward job specialties can be developed or changed, which can affect views on training later in the process.[6] GTEP recruits are already classified, but even specialties that are nominally 100-percent GTEP pick up some non-GTEP individuals during the classification process. Specialties that are less than 100-percent GTEP (such as AGE) get the remainder of their trainees through the classification process at BMT.

At the end of BMT, students head to initial skills training, consisting of at least one technical training course. The training pipeline might require courses at multiple locations over an extended period of time. All of the courses evaluated in this research were relatively long compared with the overall average for a technical training graduate. All except AGE, operations intelligence, and network intelligence required multiple courses to graduate.[7] After graduating and earning 3-level qualification (awarded apprentice status), the airman travels to his or her first duty station to continue training on the job.

In our research, we reviewed the entire process from recruiting to BMT to classification to technical training. The majority of the focus was on technical training.

Of particular interest during the technical training process turned out to be the phase program, a system of gradually decreasing restrictions after BMT. In brief, the program's goal is to transition a trainee from BMT graduate to a motivated, self-disciplined, and fit airman by progression through phases of greater responsibility and freedoms (Air Education and Training Command [AETC] Instruction 36-2216). At the time of the research, airmen spent at least 185 days (usually all of their training) in the phase program before becoming eligible for phase grad (14 days in phase I, 21 days in phase II, and 150 days in phase III).

Analysis Approach

Some studies over the past few years have focused on the high attrition rates in these specialties and others.[8] To avoid duplicating work, we reviewed the extant literature and added two pieces that we saw as valuable: a qualitative piece based on interviews and site visits and a quantitative piece based on training and personnel databases.

In the qualitative effort, we focused on attrition causes by visiting each of the training bases to talk to students and instructors about reasons for the high attrition. The interview and focus-group results also provided ideas for further investigation.

In the quantitative effort, we focused on better selection criteria by combining the training and personnel databases to examine the relationship between personnel data and training outcomes (success or failure and final grade) in these specialties.

---

5 Appendix C provides an explanation of the various exercises in the PAST.

6 BMT is designed to change attitudes, values, and behavior. Interviews suggested that negative attitudes toward a job could develop, depending on the training instructor's (TI's) attitude toward a job or the opinion (not necessarily knowledgeable) of peers.

7 The majority of specialties require only one course to graduate to 3-level qualification.

8 Some of the efforts are discussed in a later section in this chapter.

Review of Extant Literature and Existing Efforts

We found useful insights in previous work on recruiting, ATC screening tools, EOD screening tools, USAF physical-fitness studies, generational impacts, and the effect that longer educational days can have on learning, summarizing important research findings from each.

The Air Force recognizes the problem of attrition and has sponsored research on reducing attrition rates in its training programs. Table 1.3 lists the proposed and ongoing actions over the next five years, as reflected in the Air Force's Personnel Research Plan (Headquarters [HQ Air Force Personnel Directorate [AF/A1], undated).

The Force Management Liaison Office at Air Force Personnel Center (AFPC) has contracted with the Human Resources Research Organization (HumRRO) for attrition analyses. We reviewed results of the preliminary diagnostics, which looked at attrition from 2002 to 2006 by comparing attrition statistics with admission waivers for recruits.

Table 1.3
Action Items in the Air Force Personnel Research Plan

Category	Description
Database development	Develop and maintain HRRD. Document corporate memory of critical research issues.
Operational system effectiveness	Detect trends of personnel characteristics. Develop metrics for PJM. Design an APAS.
Test development and evaluation	Assess the effectiveness of AFOQT Form S. Develop efficient job-performance measures. Perform the AFOQT cyclic update. Develop new ASVAB noncognitive subtests.[a] Create new aptitude tests with incremental validity.[a]
Enlisted selection and classification	Identify aptitude requirements for individual AFSCs.[a] Assess ASVAB's validity for predicting job performance.[a] Recommend ways to prevent premature attrition.[a] Develop warfighter-selection strategies. Optimize ASVAB classification composites.[a] Evaluate the effectiveness of those who receive waivers.[a] Describe ASVAB trend analysis and accession.[a] Develop PJM, PROMIS, and PACE. Determine AFSC-specific aptitude quality distribution.[a]
Manpower and force modeling	Create performance metrics for the strategic plan. Optimize AFSC clusters through AFSC restructuring.
Officer selection and classification	Describe test-score trends of officer accessions.[a] Evaluate the validity of the SDI+.[a] Optimize the officer-assignment algorithm. Develop officership measures.[a] Assess AFOQT validity and composite structure.[a]
Aircrew selection	Revalidate PCSM components.[a] Develop and validate TBAS II.[a]

SOURCE: AF/A1 (undated).

NOTE: HRRD = Human Resource Research Databank. PJM = person–job match. APAS = Advanced Personnel Acquisition System. AFOQT = Air Force Officer Qualifying Test. ASVAB = Armed Services Vocational Aptitude Battery. PROMIS = Procurement Management Information System. PACE = processing and classification of enlistees. SDI+ = self-description inventory. PCSM = Pilot Candidate Selection Method. TBAS = Test of Basic Aviation Skills.

[a] This item is aimed at or will have an effect of reducing attrition in training.

AETC began an effort to evaluate a noncognitive tool, the Emotional Quotient Inventory (EQ-i), an Internet-based emotional-intelligence assessment. AETC is investigating the value of the EQ-i as a predictor of attrition for ATC, CCT, PJ, EOD, and pilots. We did not have enough EQ-i data to evaluate the tool's effectiveness.

The enlisted-aircrew career field manager (CFM) expressed concerns with training attrition and on-the-job attrition. As a result, the AETC Studies and Analysis Squadron (AETC SAS) and AFPC are evaluating screening tools, including the SDI+ (a variant of the Big Five Personality Measure) and the TBAS.[9]

The 720th Special Tactics Group at Hurlburt Field has been researching the attrition problem in combat controllers. The work was ongoing at the time we talked with the principal researchers. The focus is on both pipeline attrition and posttraining retention. One early finding is that the process of constantly entering a new phase of training in a new environment and with new instructors breaks continuity in training.

The U.S. Air Force School of Aerospace Medicine (USAFSAM) is investigating attrition concerns in gunship sensor operator and Predator sensor operator training. It costs $1.1 million to train each gunship sensor operator over 105 training days, and attrition is at 40 percent. High costs are attributed to simulation training, actual flight training, and munitions. Unexpectedly high attrition levels in the pipeline leads to understaffing in the field and, therefore, longer deployment for current sensor operators. USAFSAM recommends a new screening process to decrease attrition.

Finally, the Air Force Audit Agency (AFAA) recently concluded research on attrition in the PJ, CCT, tactical air control party (TACP), and EOD career fields. It found that "procedures were not in place to collect consistent H2F [hard-to-fill] career field elimination data for analysis from all relevant data systems" (AFAA, 2008, p. 4). Also, "career field managers (CFMs) did not provide potential recruits sufficient H2F career field information before enlistment or career counseling during BMT" (AFAA, 2008, p. 4). Our research had similar findings.

Interviews and Site Visits

We visited seven locations and interviewed 55 instructors (military training leaders [MTLs] and classroom) and 154 students. Table 1.4 shows a breakout by location.

The purpose of the interviews and focus groups was to collect firsthand perspectives from instructors and students about the possible causes of attrition and to explore causes that might not be identified by the analysis of existing archival data. Students were also administered a questionnaire designed to assess trainee perceptions of what determines success and failure for each of the nine careers (e.g., strength, intelligence). A summary of the questionnaire results is presented in Appendixes B and C.

In all but two cases, we interviewed instructors individually to ensure that each instructor's responses would not be influenced by the responses of other instructors and that no

---

[9] The plus sign in SDI+ is to indicate that two additional factors (covered by 80 items) have been added to the Big Five (five broad domains used to describe personality: openness, conscientiousness, extraversion, agreeableness, and neuroticism) for operational use within the military. These two factors are service orientation and team orientation. These factors measure the degree to which a person is likely to make the service a career (i.e., for long-term return-on-investment analyses) and a person's ability to accept differing roles under changing conditions (i.e., support evolving mission statements with fewer and fewer personnel). So, in the SDI+, the mnemonic for remembering the factors is "OCEAN ST" for openness, conscientiousness, extroversion, agreeableness, neuroticism, service, and team (Weissmuller, 2006, p. 3).

Table 1.4
Number of Interviews by Site and Specialty

		Interviews	
Location	**AFSC**	**Instructors**	**Students**
Lackland AFB	EOD, CCT, PJ	8	31
Sheppard AFB	AGE	5	10
Eglin AFB	EOD	6	11
Keesler AFB	ATC, CCT	8	15
Kirtland AFB	PJ	6	8
Goodfellow AFB	Intelligence (operations, network)	12	27
DLIFLC Monterey	Linguists	8	52
Total		53	154

NOTE: AFB = Air Force base. DLIFLC = Defense Language Institute Foreign Language Center.

instructor could inadvertently dominate the conversation. For all students, interviews were conducted in a focus-group format consisting of approximately four to six students at a time. Instructors selected the students randomly, except for selecting a mix of race, gender, and grade.

We used a semistructured process for both the interviews and focus groups. Interview and focus-group protocols (described in Appendix A) were used to guide the discussions. In general, there was much agreement between instructors and between students as to causes and potential remedies. We doubt collusion in this regard; rather, we suspect that this is a frequent water-cooler topic among instructors and a dorm-room discussion item among students.

Database Analysis

The focus of the database analysis was to evaluate selection criteria that might relate to attrition. We used databases from the Air Force Recruiting Service (AFRS), AFPC, AETC, Occupational Measurement Squadron, and data we developed at RAND. For privacy purposes, the data were first stripped of social-security identifiers and replaced with a unique RAND identifier in a closed environment (no external Internet connections). We then merged the various databases using the unique RAND identifier.

The AFPC HRRD combines recruiting data with some personnel data and training data. The other AFPC data (an active enlisted master personnel extract [AAE] file, i.e., an extract of the enlisted master personnel file) contain all the personnel data on each individual. The AETC data (the Technical Training Management System [TTMS] file) contains training information, test scores, and attrition and washback reasons.[10]

When a student is eliminated from a schoolhouse training program, the Air Force prepares a document called a record of administrative training action, AETC Form 125A, which indicates the reason for attrition, the commanding officer's recommended course of action for

---

[10] We also attempted to use Occupational Measurement Squadron data from a recently administered test, EQ-i. Unfortunately, the data, at the time, were too sparse to be useful.

the student (e.g., reclassification, separation), and other relevant details. The data we developed are taken from individual Form 125A packages of individuals who failed training.

We used advanced regression techniques to identify correlates of failure and success in each of the specialties investigated. We used logistic regression, a tobit regression, and special goodness-of-fit tests (all explained in this section).

Given the essentially binary outcome of success or failure in the training pipeline, attrition was modeled as a logistic regression of the form

$$\frac{P_i(pass)}{1-P_i(pass)} = \exp\left(\sum_{i=1}^{n} \beta_i X_i\right).$$

The covariates included in the model were age at the time of training, physical factors, and ASVAB subtest scores. We started out with more than 300 variables, but most of the data showed little sensitivity to success. In consideration of the fact that CCT and PJ are elite, physically demanding career fields, we included height, weight, body mass index (BMI), systolic and diastolic blood pressure, and an indicator for whether the airman was able to lift 100 pounds or more at the Military Entrance Processing Station (MEPS) physical. Results are found in Appendix F.

Model goodness of fit was evaluated on the basis of McFadden's pseudo R^2 and on the result of the Hosmer-Lemeshow test, which compares the number of observed and expected cases of success and failure within each decile of the predicted probabilities. A significant chi-square statistic indicates model misspecification, such as nonlinearity in the right-hand variables or omitted-variable bias. The sensitivity and specificity of the logistic regression model was assessed by setting the threshold predicted probability for success to 0.5 and reporting the proportions of correctly classified observations. The results are found in Appendix F.

An effort was made to predict final grade in the training course with the same set of covariates, using a tobit regression with a lower limit of 70 and an upper limit of 100 for the lowest and highest possible passing scores, respectively. The tobit is the proper tool when data are artificially bounded, as they are in this case at 70 and 100. The tobit regression coefficients do not have a straightforward interpretation as effect sizes (i.e., the amount of increase in final grade per unit change in the covariate); therefore, we created a percentage-change measure on final score. Results are found in Appendix F.

The resulting combined database contained more than 300 variables. After a considerable amount of work, we used 17 variables in our correlations (see Appendix F). The remaining variables were not usefully correlated to success and failure or to final grade. There were a lot of missing data and a lot of collinear data. The final 17 variables included nine ASVAB subtest scores, which are already used in the initial cutoff of recruits for particular AFSCs. The fact that these ASVAB subtests still have value suggests that the use of four ASVAB composite scores (mechanical, administrative, general, and electrical [MAGE]) might not be sufficiently unique to select individuals for today's vocational skills.

Organization of the Report

It became apparent during our research that, although there were some cross-cutting issues, every career field had unique issues and concerns that formed part of the attrition question. Consequently, Chapters Two through Nine address issues specific to the individual career fields. Chapter Ten discusses issues that cut across all nine career fields. Chapter Eleven concludes with our final thoughts.

Appendix A contains the interview protocol for both students and instructors. Appendix B is the student questionnaire, and Appendix C contains the results. Appendix D contains descriptions of the ASVAB, PAST, and Federal Aviation Administration (FAA) Air Traffic Selection and Training (AT-SAT) tests. Appendix E contains a more detailed discussion of trainees' perceptions of injustice. Appendix F contains a longer discussion of the analytical approach and a detailed look at the quantitative results used to evaluate selection criteria. Appendix G describes the mathematical model used to evaluate the effectiveness of increasing washbacks versus failing the student for the AGE AFSC. Last, Appendix H compares the length of technical training programs with the attrition rates.

CHAPTER TWO

Air Traffic Control, AFSC 1C1X1

Overview

ATCs coordinate en-route and terminal air traffic by use of visual, radar, and nonradar means. As they gain seniority, they supervise and manage ATC facilities (Air Force Manual [AFMAN] 36-2108, 2004).

ATC training is 72 technical training days long. ATC training is managed by the 334th Training Squadron (TRS) at Keesler AFB in Mississippi. The course pipeline, as depicted in Figure 2.1, utilizes one training location and two courses for technical training. The 72 technical training days translate to approximately 78 days in the pipeline. The recruiting service seeks to fill 100 percent of the seats through GTEP enlistments. An ATC graduate costs $39,000 to produce.[1]

Figure 2.1
Air Traffic Control, AFSC 1C1X1, Training Pipeline

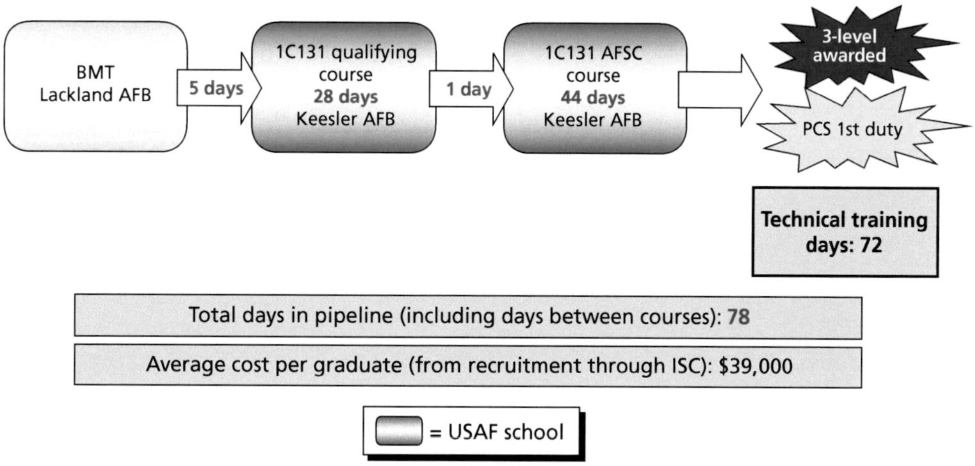

SOURCE: AETC Directorate of Intelligence, Operations, and Support (AETC/A3) (2007).
NOTE: PCS = permanent change of station. ISC = initial skills course.

RAND *TR955-2.1*

[1] Factors are based on variable costs only and are made up of the following: cost per graduate for training courses required for specific AFSCs at the basic skill level, acquisition costs (including the costs of recruiting, initial travel, and initial clothing issued), the cost of basic training at the Air Force Military Training Center (enlisted only), student pay and allowances, and time in transit and waiting for class to begin.

Candidates for the ATC AFSC are required to achieve a score of at least 55 on the general (G) composite and 55 on the mechanical (M) composite of the ASVAB test. The G composite combines evaluations of verbal expression and arithmetic reasoning. In addition, candidates must have vision correctable to 20/20 and demonstrate that they can speak English clearly.

Table 2.1 lists the attrition rates by course. At the time of our visit to Keesler, ATC trainees took the fundamentals course and then either radar or tower. In FY 2009, AETC combined radar and tower into one course.

Notable Factors

In our interviews and research on ATC issues, one unique issue stood out: the need for an ATC screener.

Screening individuals and predicting success in ATC has been a subject of research since at least the 1960s. A wide range of personal characteristics has been studied as possible bases for screening, including both cognitive and noncognitive characteristics. Currently, FAA employs an eight-test AT-SAT battery. Ramos, Heil, and Manning (2001) describe the tests and give a detailed account of the research processes that led to FAA's adoption of the battery. The thoroughness of this work and the adoption by FAA probably accounts for the apparent sparseness of more-recent research. Brief sketches of the eight tests are in Appendix D.

Carretta and King (2008) address success in USAF ATC training. They compared successful graduation from the ATC fundamentals course, grades from written tests in the course, and scores on the FAA certified tower operator test with components of the ASVAB and FAA AT-SAT tests. The subjects were 448 enlisted USAF students who had been accepted for the ATC course (as required by the design of the experiment). Thus, the researchers had a range-restricted sample, and they discuss corrections for it. Carretta and King (2008) found the general composite of the ASVAB to be a "strong predictor of academic performance." The authors point out that raising the ASVAB cut score to 60 would reduce the elimination rate but also provide too few trainees. They suggest research on the effects of using alternative composites of the ASVAB test components on diversity without degrading predictive power.

Three of the FAA AT-SAT subtests "showed potential for prediction of USAF training performance." These are angles, letter factory, and air traffic scenarios. The researchers point out, however, that each of the latter two tests takes more than an hour and a half to administer. Although the authors suggest conducting research on shortened forms of these tests, administering them would require approval for MEPS administration. Nevertheless, the likely benefit

Table 2.1
Historical Attrition Rates for Air Traffic Control by Course (%)

Course and Version	FY 2002	FY 2003	FY 2004	FY 2005	FY 2006	FY 2007	FY 2008
Fundamentals 00FB	9.8	9.6	8.9	8.2	5.8	9.5	6.0
Radar 00RB	14.4	17.8	10.7	14.3	30.5	27.0	19.1
Tower 00TB	7.1	9.7	10.8	20.9	22.2	26.0	17.0

SOURCE: AETC training database, 2009.

NOTE: After the merger of radar and tower in 2009, the elimination rates for 2009 were 27 percent actual and 27 percent programmed. Rates through August 2010 were 17 percent actual and 27 percent programmed.

of increased training success resulting from the adoption of some versions of these additional screening tests should outweigh the costs of administering the tests prior to directing students into the ATC fundamentals course.[2]

Data Analysis

In our analysis of the data, we examined Forms 125A of attrited students. Forms 125A from the ATC specialty were collected for the period between October 2007 and April 2008. Almost all attrition from the ATC career field was due to academic deficiency, as Table 2.2 shows. Medical issues, disciplinary problems, and family issues were contributing factors in 10 percent of the cases.

Students failed various combinations of blocks I, II, and III and the progress checks in between (radar and tower). One Form 125A noted that the airman's academic difficulties might be related to the recent demise of his mother. In two cases, students reported feeling highly stressed and anxious from the training program. Both were referred to life-skills counseling, and one was eventually discharged for depression. In two other cases, the students told instructors they did not have the desire or the aptitude to succeed in the ATC career field.

The quantitative analysis did not provide any factors that would be useful in selecting individuals for success in ATC (see Table F.1 in Appendix F). We did not have data to evaluate the effect that personnel factors could have on final grade.

Summary

We recommend the use of a screener, such as the FAA test, for ATC selection. There are indications that the FAA test would likely yield better results than further restricting pool size by increasing the required ASVAB score. The FAA test would require an hour and a half to

Table 2.2
Causes of Attrition for Air Traffic Control Between October 2007 and April 2008 (n = 48)

	Attrition Rate (%)	
Cause	**Listed Reason**	**Contributing Factor**
Academic	93.8	
Medical issue	6.2	4.2
Physical	0	
Voluntary (SIE)	0	
Discipline	0	4.2
Family issue	0	2.1

NOTE: SIE = self-initiated elimination.

[2] There is an administrative cost to implementing and administering an FAA-like screening test. We did not evaluate that cost.

administer, although further research might yield a shorter instrument. We recommend testing at the recruiter's office (where the PAST is conducted) versus testing at the MEPS due to the lack of additional time for other tests. The test will require a proctor to maintain the security of the test.

The quantitative results were inconclusive in identifying factors of success from the currently measured data in the USAF personnel system.

Recent Air Force Actions

At the time of the draft report (October 2010), the Air Staff was collecting data for the EQ-i test to be used as a prescreener for ATC and other battlefield-airman career fields (2AF, 2010). Additionally, the training group had requested funds to use the Air Traffic Control Intelligent Communications Environment product to help focus on a student's phraseology early on in training, with the goal of improving attrition rates (2AF, 2010). The Air Force is working on a shortened version of the FAA test and is planning validity testing in the future.

CHAPTER THREE

Combat Control, AFSC 1C2X1

Overview

A CCT provides command, control, communications, intelligence, surveillance, and reconnaissance to assist, control, and enable the application of lethal and nonlethal airpower. He provides terminal control and uses visual and electronic aids to control the airhead in the area of responsibility. Finally, he performs tactical-level surveillance and reconnaissance functions as part of the effort to build the common operating picture (AFPC, 2010).

CCT training has 175 days of instruction. CCT training is managed by the 342 TRS at Lackland AFB in Texas. The course pipeline, shown in Figure 3.1, includes six courses at five training locations for technical training.[1] The 175 technical training days translate to approximately nine months. CCT training requires an ASVAB general composite score of 44. Recruiting seeks to fill 100 percent of the jobs through GTEP enlistments. A CCT graduate costs $98,000 to produce.[2]

Table 3.1 shows the historical rates of attrition in CCT courses. CCT attrition is spread out throughout the entire training pipeline, unlike other physically demanding specialties, such as PJ, which has most of the attrition up front.

Notable Factors

In the interviews, two factors were notable and unusual, applying only to the CCT and PJ career fields. The factors were (1) an observed effect of officers assigned to training with the enlisted and (2) the importance of physical training prior to starting the school (see Table 3.2).

Effect of Having Officers on the Team

In the interviews, many instructors reported that having an officer on the team significantly reduced attrition. The instructors specifically mentioned a case in which three officers were in one class and then none in the next few classes. They felt that spreading officer trainees out among the training classes could have a positive impact on attrition.

[1] In 2009, the two Keesler courses were combined into one course.

[2] Factors are based on variable costs only and are made up of the following: cost per graduate for training courses required for specific AFSCs at the basic skill level, acquisition costs (including the costs of recruiting, initial travel, and initial clothing issued), the cost of basic training at the Air Force Military Training Center (enlisted only), student pay and allowances, and time in transit and waiting for class to begin.

Figure 3.1
Combat Control, AFSC 1C2X1, Training Pipeline

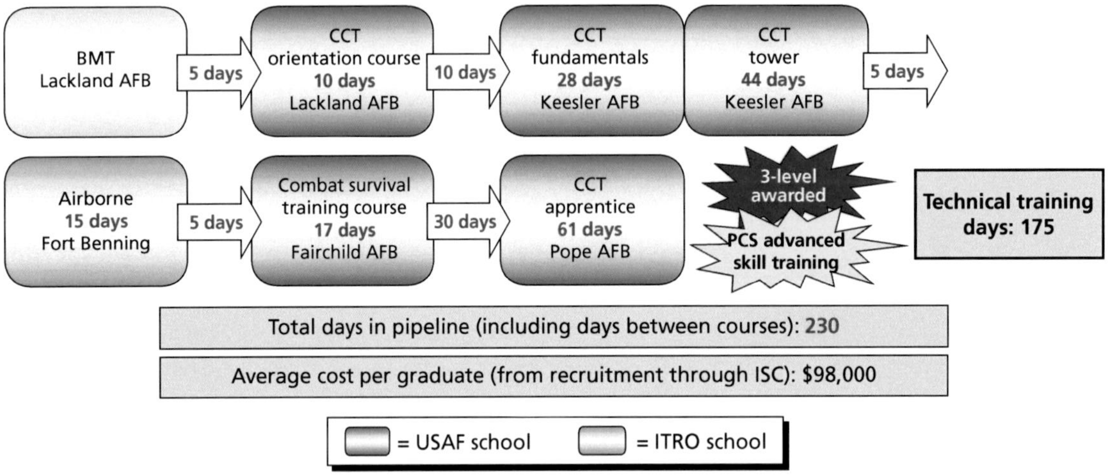

SOURCE: AETC/A3 (2007).
NOTE: ITRO = interservice training organization.
RAND *TR955-3.1*

Table 3.1
Historical Attrition Rates by Course for Combat Control (%)

Course and Version	FY 2002	FY 2003	FY 2004	FY 2005	FY 2006	FY 2007	FY 2008
Orientation 002	11.3	19.3	8.2	20.2	57.1		
Orientation 0C0A					32.6	49.4	46.2
ATC fundamentals 00FC[a]	13.7	12.6	13.7	23.5	24.6	28.7	37.6
ATC tower 00TB	5.8	12.4	25.2	18.8	13.9	35.1	26.0
Apprentice 000	35.6	27.0	27.8	34.5	0.0		
Apprentice 0C1A				0.0	27.9	40.0	41.7
Apprentice 0C1B							7.1

SOURCE: TTMS.

NOTE: This list does not include airborne or survival training courses in the CCT pipeline. An empty cell indicates that that version of the course was not used or did not exist.

[a] ATC fundamentals course attrition is higher for CCT than for ATC (see Table 2.1 in Chapter Two) due to the additional physical requirements (first two hours of the day) for CCT training.

Available data are sparse—only 15 class observations (169 students). Officer presence is statistically significant and raises the graduation rate by an average of 19.9 percentage points. The data suggest that, at a minimum, it would be worth a test of aligning at least one officer per class.

Physical Fitness

See the discussion in Chapter Eight.

Table 3.2
Impact That Having Officers Present on Teams Can Have on Combat Control Attrition

Year	Team	Total	Officer Present	Final Pass	Final Fail	Percentage Graduating
2006	1	14	Yes	14	0	100
	2	10	No	8	2	80
	3	12	Yes	8	4	67
	4	8	Yes	7	1	88
	5	11	No	8	3	73
	6	13	No	8	5	62
	7	11	No	6	5	55
2007	1	14	No	9	5	64
	2	12	Yes	8	4	67
	3	10	Yes	6	4	60
	4	12	No	6	6	50
	5	10	No	5	5	50
	6	11	No	3	8	27
	7	9	No	4	5	44
	8	12	No	7	5	58

NOTE: The USAF does not officially collect these data. The 334 TRS provided the data.

Data Analysis

According to 121 Forms 125A we examined, voluntary disenrollment accounted for two-thirds of all attrition from the CCT pipeline at Lackland and Keesler AFBs from October 2006 to April 2008. Self-elimination was a bigger fraction of attrition in the orientation course (75.6 percent), but it remained high even in the later courses (59.0 percent), in which the right was sometimes exercised before the first day of class. The remaining attrition was mainly due to medical issues and physical deficiency (see Table 3.3). As in the PJ specialty, the physical demands of CCT training led to injuries and to voluntary disenrollment arising from insufficient fitness, aversion to the water training, and, to a lesser extent, aversion to the grass and guerrilla drills.[3]

A common remark on Forms 125A from the orientation course was "CCT is definitely not for him." Also recorded in the justification were such comments as "airman doesn't want to be here." One case of voluntary disenrollment indicated that the recruiter had placed the recruit in CCT against his wishes.

[3] Grass drills are exercise movements that feature rapid changes in body position (usually on the grass). Guerrilla drills are used to improve agility, endurance, and, to some degree, muscular strength, in individual and partner exercises. These drills require soldiers to change their positions quickly and do various basic skills while moving forward.

Table 3.3
Causes of Attrition for Combat Control Between October 2006 and April 2008 (n = 121)

	Attrition Rate (%)	
Cause	Listed Reason	Contributing Factor
Voluntary (SIE)	66.7	
Medical issue	14.0	5.0
Physical	11.6	21.5
Academic	4.1	
Discipline	1.7	1.6
Family issue	1.7	2.5
Hydrophobia		9.1

Interviews with instructors at Keesler indicated that many SIEs were also academic—the student realized that he was struggling with the ATC material and therefore did not feel that CCT was a good career field for him. The academically induced SIEs could be reduced through a use of an ATC screener, similar to the one recommended for the ATC career field.

The larger data analysis did not show a large improvement (4-percent explanation in variance) in selecting individuals for success in CCT (see Table F.2 in Appendix F). Two factors, the ability to lift more than 100 pounds and age at entry (younger is better), were unsurprisingly significant. Additionally, predicting final grade had little effect (see Table F.3 in Appendix F).

Exemplars

See the exemplar discussion of the phase program in our analysis of the PJ specialty in Chapter Eight.

Summary

SIE accounted for 75 percent of all attrition from the CCT pipeline at Lackland and Keesler AFBs between October 2006 and April 2008. We recommend that more information be provided to recruits, explaining what the job entails, as well as the demands of the training. Students interviewed suggested videos of the training and of what life is like in the field.

There is a statistical correlation between officers training with the team and a lower attrition rate (20-percent improvement). We recommend a small test of spacing officer trainees out, at least one to a class.

Another cause of CCT attrition is inadequate physical fitness. We discuss our observations about physical fitness and the Fit Flight concept under PJ findings later in the report.

Although weak statistically, the data analysis does support previous findings by indicating a correlation with ability to lift more than 100 pounds. Also, age is negatively correlated with success. We believe that other factors (motivational and psychological factors, measurable by the EQ-i or alternative measure) might provide better selection criteria.

Recent Air Force Actions

At the time of the draft report (October 2010), the Air Force was collecting data for the EQ-i test to be used as a prescreener for CCT (2AF, 2010). Additionally, as mentioned in the discussion of ATC, the training group had requested funds to use the Air Traffic Control Intelligent Communications Environment product to help focus on student's phraseology early on in training, with the goal of reducing attrition and washback rates (2AF, 2010). Actions affecting ATC will affect CCT outcomes as well.

CHAPTER FOUR

Operations Intelligence, AFSC 1N0X1

Overview

The operations intelligence specialty performs and manages intelligence activities and functions, including developing, evaluating, and providing intelligence information. The job also requires instructing aircrews on collection and reporting requirements (AFMAN 36-2108, 2004).

Operations intelligence training requires 110 instructional days and is taught by the 315 TRS at Goodfellow AFB in Texas. The course pipeline, shown in Figure 4.1, utilizes one technical training location. The 110 technical training days translate to 138 calendar days—approximately five months. Candidates for the operations intelligence AFSC are required to achieve a score of at least 57 on the general composite of the ASVAB test. The general composite combines evaluations of verbal expression and arithmetic reasoning. Each graduate costs $40,000 to train.[1]

Table 4.1 lists the attrition rates by course.

Attrition has ranged from a low of 12.2 percent in 2005 to a high of 33.7 percent more recently in 2007. The temporary spike in 2007 was caused by curriculum changes.

Figure 4.1
Operations Intelligence, AFSC 1N0X1, Training Pipeline

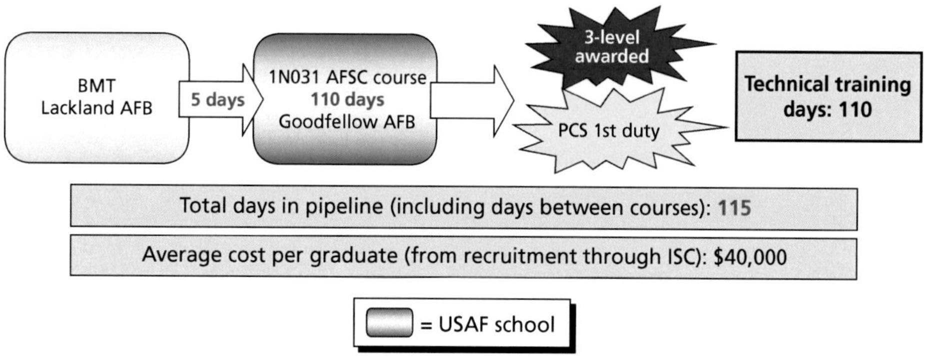

SOURCE: AETC/A3 (2007).

RAND *TR955-4.1*

[1] Factors are based on variable costs only and are made up of the following: cost per graduate for training courses required for specific AFSCs at the basic skill level, acquisition costs (including the costs of recruiting, initial travel, and initial clothing issued), the cost of basic training at the Air Force Military Training Center (enlisted only), student pay and allowances, and time in transit and waiting for class to begin.

Table 4.1
Historical Attrition Rates for Operations Intelligence (%)

Course	FY 2002	FY 2003	FY 2004	FY 2005	FY 2006	FY 2007	FY 2008
Operations intelligence	13.9	16.9	16.2	12.2	18.7	33.7	17.9

SOURCE: TTMS.

Notable Factors

In our interviews, we noted five factors that were more characteristic of the operations intelligence career field than of others: concern about the frequent noninstructional time, the quality of some instructors, a misleading job description, the need for certain skills, and questions about why a similar career field had considerably less attrition.

Instructional Issues

The most frequent complaint concerned the amount of unstructured time in the training day. Trainees consistently claimed that the instructors routinely left the classroom and that this designated study time quickly devolved into "decompression time." A prior-service (PS) trainee alleged that he kept tabs on the amount of time the instructor was in the classroom for five consecutive days and said that the instructor was present only 12 hours of the 40-hour training week.[2] Given the length of time devoted to instruction, the refrain from trainees was that the pipeline could be made shorter, by as much as several days per block. To the extent that AFSCs with longer pipelines tend to have higher attrition rates, prolonging a training course with unsupervised—and reportedly unproductive—study periods could exacerbate attrition.

Instructional Quality

Trainees were also critical of the quality of instruction. According to some, presentation of the material is rather dull and perfunctory, adhering closely to the readings, with few personal anecdotes or discussion sessions to enrich understanding. These instructional deficiencies were attributed to lack of dedication or operational experience. There was support for this contention even among instructors whom we interviewed. One PS trainee pointed out that two of the instructors in operations intelligence had never left the school (had no actual experience in the job).[3]

Job Description

Another issue that surfaced repeatedly was the disparity between reality and how the operations intelligence career field is described to recruits. Several trainees and a few instructors recalled reading or being told that operations intelligence was "like being James Bond," which they felt in hindsight was an obvious attempt to romanticize the more mundane realities of military intelligence work. Although such salesmanship might be necessary to meet USAF quotas, it could also attract the wrong kind of people to the field and disappoint others by

[2] This individual assertion was validated independently by other interview groups.

[3] This assertion was validated by instructors.

creating unrealistic expectations, thereby potentially contributing to both involuntary and voluntary attrition. Trainees and instructors confirmed that some academic failures are deliberate schemes to be reclassified.

Need for Certain Skills

In considering what skills the career field actually requires, interviewees candidly offered suggestions on possible screening tests for entry into operations intelligence. One instructor simply advocated a "much higher" ASVAB score requirement for the schoolhouse, but the aspects of the training identified as most difficult by students and instructors alike were not academic per se. The two most common suggestions were tests of ability and propensity for memorizing large amounts of arcane information and an oral exam to identify airmen with public-speaking ability.

Instead of a screener, another option would be self-paced instruction. Students memorize information at different speeds, a fact reaffirmed in the interviews. Some interviewees were bored, but some could not keep up. We speculate that self-pacing would decrease the overall average course length by decreasing washbacks for those who cannot keep up and by reducing the graduation date for those who can learn more quickly.

Comparison Between Operations Intelligence and Imagery Analysis

TRS leadership suggested comparing operations intelligence (AFSC 1N0XX) and imagery analysis (AFSC 1N1XX) because they are similar in length and qualifications but with quite different attrition rates (see Table 4.2).

The two specialties are similar in length—operations intelligence is one week shorter, 29.9 weeks versus 31.7 weeks. Operations intelligence has a lower ASVAB cutoff score for general, 57 versus 66, which suggests a slightly lower-quality recruit.[4] The cutoff score should not make that drastic of a difference, although it would have some effect. The educational requirements for operations intelligence are broad (desirable courses are speech, journalism, geography, modern world history, statistics, algebra, geometry, and trigonometry), while imagery

Table 4.2
Historical Attrition and Washback Rates for Operations Intelligence (AFSC 1N0XX) and Imagery Analysis (AFSC 1N1XX) (%)

Course	FY 2002	FY 2003	FY 2004	FY 2005	FY 2006	FY 2007	FY 2008
Operations intelligence attrition	13.9	16.9	16.2	12.2	18.7	33.7	17.9
Operations intelligence washbacks	40.9	36.6	35.9	46.6	48.4	59.3	—
Imagery analysis attrition	8.1	9.8	4.2	10.0	3.2	3.6	4.2
Imagery analysis washbacks	32.2	26.2	15.4	35.7	18.0	18.5	—

SOURCE: TTMS.

NOTE: Data are incomplete for washbacks in 2008.

4 We cannot cite any empirical evidence, but it seems unreasonable that a 200-percent increase in attrition should result from a 10-percent decrease in ASVAB requirements.

analysis is more specific and technical in focus (desirable courses are mathematics, advanced English, and computer applications).

A look at the knowledge requirements for the specialties (AFMAN 36-2108, 2004) suggests another difference (Table 4.3).

A much broader knowledge base is required for the operations intelligence student, suggesting more memorization. Imagery analysis students, while also requiring memorization, spend a significant amount of time in understanding how to interpret imagery. Interpretation probably does require a higher ASVAB but also makes the training more interesting than memorizing large amounts of information, and the more-interesting training might account for most of the difference in the attrition rates.

Data Analysis

We examined Forms 125A from the operations intelligence course generated between mid-2007 and the end of 2008. Roughly four-fifths of attrition from operations intelligence training was for academic deficiency, as Table 4.4 shows. Disciplinary problems, medical issues, and one security-clearance issue accounted for the remainder of the official reasons for disenrollment. More-minor disciplinary problems, medical issues, and family issues were contributing factors in some cases.

More than half of the attrition (55 percent) from operations intelligence occurred in the first 36 training days of the 110-day pipeline. Academic problems typically surfaced in the first few blocks of training. When academic disenrollment occurred in blocks 10–15, there was usually a contributing factor. Attrition ascribed to disciplinary, medical, or security issues could occur at virtually any point in the training. As in the network intelligence career field

Table 4.3
Comparison of Knowledge Requirements Between the Operations Intelligence Specialty and the Imagery Analysis Specialty

Operations Intelligence	Imagery Analysis
3.1. Knowledge . . . intelligence organizations and systems; collection and reporting . . . , procedures, and methods; intelligence information sources; . . . identifying, . . . analyzing information; geographical and cultural aspects of foreign countries; current military capabilities and employment tactics of potential enemy . . . weapon systems; special operations; procedures for acquiring, updating, and maintaining intelligence documents, maps, and charts, . . . graphic, oral, and written intelligence information presentation; target planning and materials; target folder construction . . . ; capabilities and application of automated data handling and management systems; security classification marking and control; US sensor systems, regional physical characteristics relative to radar significance; methods of verifying target intelligence information derived from imagery; basic electromagnetic theory; computerized systems supporting target intelligence and mission planning systems; digital terrain and feature data bases; . . . principles of precise positioning systems and targeting and weaponeering.	3.1. Knowledge . . . basic and advanced imagery interpretation principles, techniques, and procedures for imagery exploitation, reports, and presentations; . . . imagery intelligence collection systems and procedures; techniques of collating, analyzing, and evaluating imagery intelligence; use of maps, charts, grid systems, and interpreting equipment to solve imagery intelligence problems; mosaic construction; intelligence reference materials; fundamental mensuration techniques; distribution of imagery intelligence; . . . uses of target and imagery intelligence data; production of imagery related target materials; and security controls, ...

SOURCE: AFMAN 36-2108, 2004.

Table 4.4
Causes of Attrition for Operations Intelligence for 2007 and 2008 (n = 67)

	Attrition Rate (%)	
Cause	**Listed Reason**	**Contributing Factor**
Academic	80.6	
Discipline	10.4	10.1
Medical issue	7.5	4.3
Security issue	1.5	1.5
Family issue		8.9

(discussed later in this report), the Forms 125A of several airmen in operations intelligence noted a medical condition of insomnia or disciplinary action for oversleeping or falling asleep during the duty day. Other disciplinary problems included cheating on an exam, illegal drug or alcohol use, and a pattern of insubordination. Medical disenrollments were for pregnancies, anger management, anxiety, and unspecified maladies. Almost all family issues involved the death or serious illness of a loved one, and the security elimination was for an unspecified off-duty incident that called the airman's character into question. A handful of eliminated trainees cited negative peer influences (a distracting roommate, overbearing PS member), being phased back just prior to an exam, and general difficulty transitioning between course and squadron commitments as contributing factors in their failures.

The larger data analysis, correlating personnel factors with success, showed a very slight potential improvement (3-percent explanation in variance) in selecting individuals for success in operations intelligence (see Table F.4 in Appendix F). Also, evaluating personnel factors that correlate with predicting a higher final grade had little effect (see Table F.5 in Appendix F).

Summary

Interviews revealed that some instructors routinely leave the students in unsupervised—and reportedly unproductive—study periods. We recommend self-pacing for this course in a supervised environment.

During the period of the research, a few instructors were first-assignment instructors. The students considered them subpar. We recommend not using instructors who lack experience in the field.

Interviews of instructors and students suggested that the job description given during recruiting is not representative and might glamorize the job. We recommend that the job description accurately reflect the job, within the constraints of security.

The training requires a lot of memorization, and the job requires public-speaking ability, suggesting that this specialty might be amenable to a screening test, possibly an oral one.

Recent Air Force Actions

At the time of the draft report (October 2010), the Air Force had pursued a course rewrite and merger of multiple intelligence courses with the goal of reducing rote memorization of facts and incorporating more application-based training. The Air Force feels that the new courses are not conducive to self-study, with the possible exception of the 1N core block of instruction (2AF, 2010).[5]

Instructor manning has improved to 80 percent due to the implementation of a new deployment system and clearer guidance to AFPC on manning requirements (2AF, 2010). The issue with first-assignment enlisted instructors has been corrected and those billets removed (2AF, 2010).

---

[5] We were not able to independently verify assertions regarding self-study, although the rationale for not requiring self-study is correct.

CHAPTER FIVE

Far East and Middle East Linguists, AFSCs 1N3X4X and 1N3X5X

Overview

Those in linguist specialties perform and supervise identification, acquisition, recording, transcribing, translating, analyzing, and reporting of assigned voice communications (AFMAN 36-2108, 2004).

In the course of the research, we interviewed ground-based linguists (AFSC 1N3XX) and airborne linguists (AFSC 1A8X1) who were taking the same courses. Although there are differences in job functions and duty locations, the language components are essentially the same. The focus of the research was the ground-based linguist, but we also report on findings related to airborne linguists.

Ground-based linguist training lasts up to 435 training days, depending on the language studied. Primary training is conducted at DLIFLC in Monterey, California, and follow-on cryptologic training is held at Goodfellow AFB in Texas (Figure 5.1). Airborne linguists complete additional physiological and survival courses that lengthen training, as shown in Figure 5.2. Including time between courses and block leave, an airman can spend close to two years in the training pipeline before reporting to his or her first duty station. Recruiting seeks to fill 100 percent of linguist training seats through GTEP enlistments. It costs approximately $160,000 to produce each linguist.[1]

Airmen are selected for linguist training based on scores in the ASVAB general composite (G > 72) and Defense Language Aptitude Battery (DLAB) scores, which determine the category of language an airman may study. Category III languages require a score of 105, and category IV languages require a 110.[2] The maximum DLAB score is 176. The language category reflects the difficulty of the language and length of training: Training for category III languages (e.g., Persian Dari, Persian Farsi, Pashto-Afghan, Urdu) is 48 weeks, while category IV language courses (e.g., Arabic, Mandarin Chinese, Korean) last 64 weeks. Course lengths include a one-week orientation.

---

1 Factors are based on variable costs only and include the following: cost per graduate for training courses required for specific AFSCs at the basic skill level, acquisition costs (including the costs of recruiting, initial travel, and initial clothing issued), the cost of basic training at the Air Force Military Training Center (enlisted only), student pay and allowances, and time in transit and waiting for class to begin.

2 These scores became effective in FY 2007; previous qualifying scores were 95 for category III and 100 for category IV. In a recent journal article, the DLIFLC commandant noted, "the average DLAB score for all incoming students in FY08 was 113—just about what it was before we raised the requirement." She also stated that, due to the increased requirement, "about 25 percent of our students are here on [DLAB] waivers, compared to 2006 (prior to the change) when only one percent required waivers" (Sandusky, 2009).

Figure 5.1
Ground-Based Cryptologic Language Analyst, AFSC 1N3XX, Training Pipeline

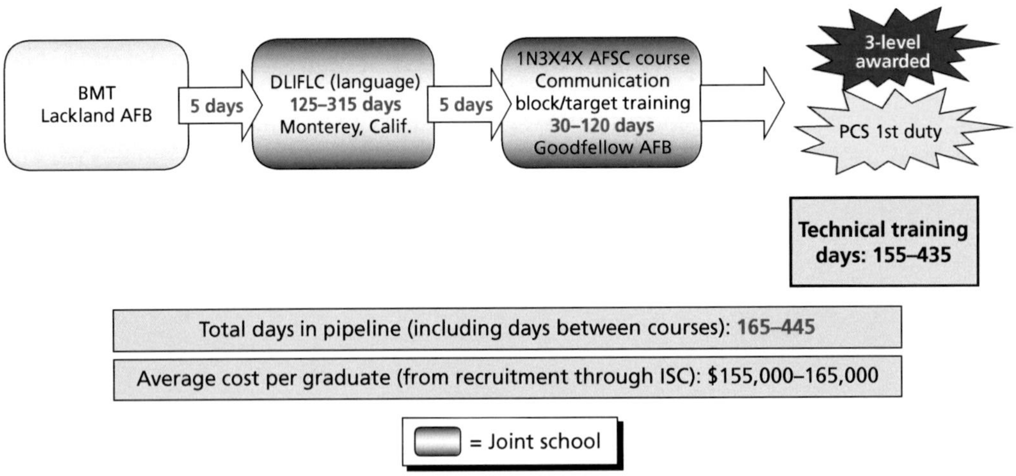

SOURCE: AETC/A3 (2007).

RAND *TR955-5.1*

Figure 5.2
Airborne Cryptologic Language Analyst, AFSC 1A8X1, Training Pipeline

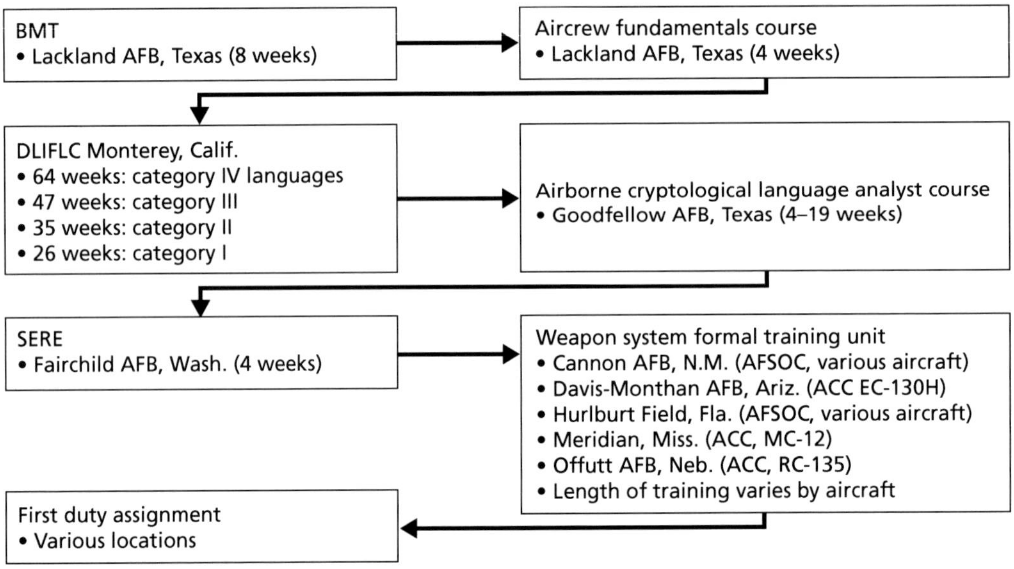

SOURCE: HQ USAF Intelligence, Surveillance and Reconnaissance Directorate (2009).
NOTE: SERE = survival, evasion, resistance, and escape. AFSOC = Air Force Special Operations Command. ACC = Air Combat Command.

RAND *TR955-5.2*

Recruits programmed for the linguist specialty enter the Air Force knowing that they will continue onto linguist training after BMT, but they do not know which language they will study. During BMT, each prospective linguist fills out a language preference form. Some we interviewed were notified of their projected language during BMT; others did not find out until arriving at DLIFLC.

Scope

DLIFLC offers training in 23 languages and associated dialects. Our research centered on Middle East and Far East languages with large numbers of USAF enlisted students. These languages, along with the number of airmen training in each during FY 2008, are as follows:

- Middle East: Arabic (356), Dari (56), Pashto (67), and Farsi (85)
- Far East: Chinese (202) and Korean (290).[3]

In order to understand the training process and reasons for success and failure, we interviewed more than 100 personnel during a four-day site visit in February 2009. These personnel included DLIFLC and TRS leadership, language teachers and students, and support personnel. Subsequently, we analyzed USAF training data and AETC Forms 125A.

Structure

Linguist training differs from the other career fields described in this report in two specific respects:

1. The Army operates DLIFLC. USAF training squadrons do not directly oversee USAF linguist trainees during the duty day.
2. DLIFLC's instructor force is made up of primarily civilian native speakers. The academic side of DLIFLC evinces a collegiate, rather than military, atmosphere. A small number of military language instructors assist in the education.

The DLIFLC commandant is an Army colonel (O-6), and the assistant commandant is a USAF colonel who is dual-hatted as the commander of the 517 Training Group (TRG), responsible for all USAF activity at DLIFLC. USAF student personnel are assigned to either the 311 TRS or the 314 TRS, depending on language of study. The Air Force represents the largest share of students at DLIFLC with approximately 1,200 students, compared with 850 Army students, 525 from the Navy, and 325 Marines.

DLIFLC's separate line of authority for academics oversees linguist trainees during the duty day. Each language school is led by a civilian dean (responsible for curriculum, academic and administrative policy, manpower, and budget) along with a field-grade (O-4 or O-5) military associate dean who monitors student progress and directs the school's military language instructor (MLI) program. It appears that the language schools and USAF squadrons maintain a good relationship; recurring meetings keep the squadrons informed of which students are succeeding and which are struggling.

Table 5.1 lists the attrition rates by course for FY 2008. Two factors should be kept in mind when assessing these rates. First, these numbers reflect only academic attrition. An additional source of overall attrition is the Defense Language Proficiency Test (DLPT), the end-of-course test that must be passed to continue in the linguist career field. Additionally, due to the length of training, a student's time at DLIFLC crosses parts of two, and occasionally three, fiscal years.

[3] From the 311 TRS and 314 TRS mission briefing, dated October 29, 2008.

Table 5.1
Fiscal Year 2008 Attrition Rates by Language

Course	Attrition Rate (%)
Mandarin Chinese	29.7
Korean	26.2
Arabic	28.1
Dari	33.3
Farsi	23.9
Pashto	19.7

SOURCE: 311 TRS and 314 TRS data.

Notable Factors

Several issues stood out during interviews and research into USAF linguist training.[4] In the following paragraphs, we discuss motivation required to succeed in language learning, TRS support, language placement, and proficiency testing.

The Importance of Motivation

When we discussed factors essential for successful language study, staff, instructors, and students all cited the importance of motivation. For example, on our written survey, 35 out of 41 students agreed (selected either "agree" or "strongly agree") with the survey statement "I'm proud to have this AFSC," and 37 agreed with the statement "It is really important that I do well in this training." These views on motivation match the opinion of language-learning expert Zoltan Dorynei (2001):

> During the lengthy and often tedious process of mastering a foreign/second language, the learner's enthusiasm, commitment, and persistence are key determinants of success or failure. . . . [W]ithout sufficient motivation, even the brightest learners are unlikely to persist long enough to attain any really useful language.

Student comments on motivation centered on making progress in the language, appreciation of teachers, and interest in cultural aspects of the course. A typical comment was that DLIFLC is "intense and interesting since we're working with native speakers." Those studying languages that have been taught at DLIFLC for some time (Chinese, Korean, and Russian) were more positive about their experiences than those studying more–recently added languages (Dari, Farsi, or Pashto). Remarks indicated that DLIFLC is going through growing pains in these languages due to operational requirements, so challenges (such as new instructors, new

[4] We interviewed more than 100 personnel at DLIFLC, including squadron leadership, DLIFLC staff, and students representing the following languages: Arabic, Chinese, Persian Dari, Persian Farsi, Korean, Pashto, Russian, and Tagalog. Interviews were conducted in a focus-group format. Each student also filled out an anonymous survey that included 37 statements (answered from strongly agree to strongly disagree) and five free-response questions. The survey is found in Appendix B. The compiled responses are in Appendix C.

materials, and the availability of MLIs) affect the program. Overall, students indicated strong support on the written survey, as 34 of 41 agreed (either "agree" or "strongly agree") with the statement "My instructors support my job of learning a second language."

Factors that decrease student motivation include the difficulty of language courses and squadron (military) obligations, such as the phase program. Students stated that they understood the need for the length and intensity of language learning, but many expressed that this aspect of the training can be wearing. Representative comments included, "DLI [the Defense Language Institute] is a marathon of academia. I would have preferred college if I wanted academia," and "At DLI it's easy to lose focus and motivation. You can get fried."

Training Squadron Support: Academic and Military Leadership

In focus-group discussion, opinions of squadron support were mixed. Airmen appreciated the reduction in military obligations associated with progression through the phase program. A few also stated that some flights supported academic efforts by reducing military obligations during finals and when classes were scheduled for the DLPT. Others said that their flights did not do this. The strongest statement of support came from an airman who stated that the squadron "was absolutely supportive—always asking what could be done to help the students." Responses to the written survey were more skeptical, as only 12 of 41 agreed with the similar statement, "My squadron supports my job of learning a second language" (nine answered "neither agree nor disagree," while 19 selected "disagree" or "strongly disagree").

Regarding the training squadrons, students most appreciated the role of their academic training advisers (ATAs). Common remarks among those we interviewed reflected that the ATAs knew the difficulty of learning a second language: "[T]hey get it because they've been there."

We did not hear of egregious MTL behavior, although students indicated that MTLs "don't know what we're going through since they haven't studied language at DLI." The other services at DLIFLC use linguists in MTL-like roles, a practice that perhaps the Air Force should consider as well.

MTLs were also seen as the face of the phase program, a unifying sore point among the trainees we met. Comments on the program were similar to those heard at other technical training bases, with the exception that DLIFLC students stressed the length of their program, inequity when compared with their BMT colleagues who were already at active units, and a perception that standards for phase graduation were not evenly applied. Interviews indicated that academic, physical fitness, and volunteer standards exceeded the minimum requirements for phase graduation set in the AETC instruction governing the phase program (AETC Instruction [AETCI] 36-2216, 2004).

We interviewed DLIFLC instructors from three large language schools. They seemed energized about their role teaching future linguists and impressed (on the whole) with the students in their classes. When we asked what caused students to fail, answers centered on motivation, ability, class chemistry, course difficulty, and inability to handle distractions. More than once, we heard teachers who grew up outside the United States comment that the U.S. education system is not producing students who know how to study. We also heard the interesting critique that U.S. students need to be convinced why language study is useful, rather than finding utility in learning itself. A final criticism that we heard from instructors and squadron personnel alike is that video games are a diversion that takes valuable study time away from linguist trainees.

MLIs in these same language schools all agreed that the current curriculum is more difficult than what they studied at DLIFLC (seven to 15 years ago) and agreed that DLIFLC is producing better linguists as a result. However, one MLI stated that "DLI is the hardest thing most of them will ever do. Some of them just give up." The MLIs provided the best insight into the washback process, stating, "[U]nless a washbacked student is very motivated and has a diagnosable weakness, they don't generally do well on the DLPT even if they make it through the course."

Language Placement

Students and squadron staff expressed frustration regarding the language placement process. As alluded to earlier, prospective linguists indicate language preferences while at BMT. It was striking that, among those we interviewed, very few were placed in languages of their choosing, including those who had background in their preferred language. Given that motivation is essential, and that studying a preferred language should increase motivation, this situation is puzzling. Further exacerbating the situation is the fact that the 311 TRS and 314 TRS do not have the authority to adjust languages among airmen. For example, if airman A is assigned to Russian but wants to study Chinese, and airman B is assigned to Chinese but wants to study Russian, a swap cannot be made at squadron level and cannot be made at all without extreme effort.

Language Proficiency Testing

Regarding the capstone DLPT, the final hurdle a DLIFLC student faces before being declared a graduate, one airman stated, "The biggest frustration is that you can make it all the way through the course and fail to qualify because of the DLPT." Linguists must score at "limited working proficiency" (level 2 or better on the Interagency Language Roundtable scoring system) to graduate as a linguist and receive the monthly proficiency bonus. Squadron statistics showed that, over the past three fiscal years, 76 percent of airmen completed the basic course of study and approximately two-thirds of these students passed the DLPT at proficient levels, meaning that the overall completion rate was near 50 percent—higher in some languages, lower in others. This pass rate among highly qualified airmen (measured by the ASVAB) causes us to question whether expectations are too high.

DoD implementation of the Defense Language Proficiency Test Five (DLPT5) compounds concern about this high-stakes test. According to DLIFLC, DLPT5 better integrates authentic materials and allows DoD to accurately measure proficiency levels. However, the assumptions behind DLPT5 take into account that the test is more difficult than earlier versions. DLIFLC's DLPT5 page of frequently asked questions includes the following response to the question, "Why are scores lower on the DLPT5 than on older DLPTs?"

> No language test is 100% accurate. So when calibrating a language test, the question is always whether the error should be in the direction of being generous or strict. Older DLPTs were calibrated so that the error would be in the direction of being generous. With DLPT5, the post-9/11 emphasis on readiness resulted in a desire to have any error be in the direction of being strict. (See DLIFLC, undated)

This change to "the direction of being strict" had greatest impact on airmen in the Arabic program. Under DLPT4, the pass rate for those completing the basic Arabic course was

87.9 percent, but that dropped to 23 percent with DLPT5.[5] The change affected other languages to a lesser extent: Spanish and Chinese pass rates also dropped from more than 90 percent to near 60 percent, and Farsi dropped from more than 90 percent to 45 percent. It should be noted that similar drops occurred when DLPT4 was implemented in the late 1980s.[6] The jolt caused by DLPT5 pass rates caused DLIFLC to reform the content of its basic language courses, as well as, in some cases, recalibrate the DLPT itself.

Regarding these final two issues—language placement and proficiency testing—we recommend that the training squadrons be allowed greater latitude to place trainees in languages; we also recommend further research regarding DLPT qualifications required for graduation, including the possibility of allowing some subproficient linguists to proceed to cryptologic training at Goodfellow AFB.[7]

Data Analysis

Forms 125A from the Middle East and Far East linguist career field were collected from 2006 through 2008. The sample consisted of 96 eliminations from Arabic, 94 eliminations from Mandarin Chinese, and 34 eliminations each from Pashto and Persian Farsi. The vast majority of attrition was due to academic deficiency, as Table 5.2 shows. The balance of attrition was classified as medical issues, disciplinary problems, and family issues. As a rule, the DLIFLC Forms 125A were not very detailed—often with a one-sentence justification for disenrollment—and most came with little or no supporting documentation. Thus, the proportion of cases with a contributing factor is necessarily a lower-bound estimate.

Within the four language schools examined, attrition due to academic deficiency ranged from 85 percent (Arabic, Chinese) to 94 percent (Persian Farsi). The mean DLAB score for academically eliminated trainees was significantly higher in the Mandarin Chinese and Arabic schools (118.6) than the Persian Farsi and Pashto schools (109) ($p < 0.01$), reflecting the higher entry requirements for Arabic and Chinese described earlier.

Table 5.2
Causes of Attrition Among Middle East and Far East Linguist Trainees Between 2006 and 2008 (n = 258)

	Attrition Rate (%)	
Cause	Listed Reason	Contributing Factor
Academic	86.8	
Discipline	5.8	3.1
Medical issue	6.6	1.9
Family issue	0.8	

[5] From the 311 TRS and 341 TRS mission briefing, dated October 29, 2008.

[6] See briefing by Lett and Dege (2001). Pass rates in Arabic, Chinese, and Korean were below 20 percent in 1989, but each grew to more than 80 percent by 2001.

[7] Cryptological training adds language instruction in military-specific terms.

Among those who were ultimately unsuccessful in these DLIFLC language programs, most failed in the pipeline, but some academic attrition was the result of failure to achieve an adequate score on the DLPT taken at the end of the training. Overall, the proportion of academic attrition due to failing DLPT scores was 29.5 percent, but, among languages, the proportion of academic attrition due to DLPT failure ranged widely, from 11.8 percent (four out of 36) in Pashto to 43.8 percent (42 out of 96) in Arabic. Two factors explain the proportion of DLPT academic failures in the Arabic school. First, the Arabic department has a policy of not academically eliminating trainees in the third term of the course. Additionally, implementation of the new Arabic DLPT (DLPT5) saw a large increase in failures. Two airmen who failed the DLPT in Arabic were reclassified into a language for which they had achieved passing DLPT scores (one into Russian, one into Spanish).

USAF success rates are similar to those of the other services. In FY 2008, 96 USAF Chinese linguists graduated, out of 202 who started the program (47.5 percent), while sister services were a combined 77 graduates out of 157 (49 percent). These students were tested with DLPT5. All services had better FY 2008 Korean rates, as the Air Force was 214 of 290 (73.7 percent), while other services were 129 of 191 (67.5 percent).[8] These rates are with the DLPT4. The Korean language began using DLPT5 in 2010.

There were minor differences in the breakdown of attrition causes between airborne (1A8X1) and ground (1N3X4/1N3X5) linguists, as shown in Table 5.3.

Disciplinary eliminations among Middle East and Far East linguists were mostly for lack of effort (LOE) in training, as opposed to lack of ability (LOA). The distinction between LOE and LOA is determined jointly by DLIFLC's academic departments and military units. The LOE justification appeared only in the Forms 125A of Chinese linguists, which by itself explains the higher proportion of disciplinary attrition in the 1N3X4 career field. Two airmen were disenrolled pending Article 15 hearings for unspecified transgressions, and another was eliminated for breaking curfew and underage drinking. Only two trainees in the sample were essentially eliminated for family issues: The Form 125A of one Chinese linguist noted that his grades deteriorated rapidly following the loss of a family member and friend, and a student in the Arabic language course was discharged from the Air Force for hardship. At least eight trainees were eliminated for pregnancies (the actual number might be higher, but the reasons for medical separation were not always provided). Curiously, seven of the eight were trainees in Middle East languages. Two airmen were reclassified from AFSC 1A8X1 to 1N3X5 for medi-

Table 5.3
Comparison of Attrition Rate Reasons Between Airborne and Nonairborne Middle East and Far East Linguist Trainees (n = 258)

	Attrition Rate (%)		
Cause	**1A8X1 (n = 102)**	**1N3X4 (n = 56)**	**1N3X5 (n = 100)**
Academic	89.2	80.4	88.0
Discipline	4.9	14.3	2.0
Medical issue	4.9	3.6	10.0
Family issue	1.0	1.8	0.0

[8] Data provided by DLIFLC Asian Language School.

cal conditions that disqualified them from flying, such as short stature and frequent migraine headaches. Since these reclassifications entail no new training, these are costless and not really part of the attrition problem.

With regard to contributing factors, there was generally insufficient information in the Form 125A file to make an inference. In a few cases of academic attrition, there were also indications of problems adapting to military life. For example, one airman in the Chinese course had received 13 341s (demerits) for failure to comply with grooming standards, and a student in the Arabic course was reprimanded for underage drinking and counseled for missing nine physical-training formations. Health problems contributed to a small number of academic attritions because the ailments caused prolonged absences.

The larger data analysis showed a very slight potential improvement (3-percent explanation in variance) in selecting individuals for success in linguist training (see Table F.6 in Appendix F). Also, predicting final grade had little effect (see Table F.7 in Appendix F).

Summary

Interviews with students and instructors, and previous research, Dornyei (2001), confirmed the importance of motivation (enthusiasm, commitment, and perseverance) to learn a foreign language. The length and intensity of the phase program are major factors in decreased motivation (see Chapter Ten for further discussion and recommendations).

Students did not feel that the MTLs, with no experience learning a foreign language, understood the difficulty of the program.

Language requirements are determined by operational needs. And not getting assigned to duty in a preferred language can also reduce an individual's motivation. Yet training squadrons did not have the authority to allow students with similar DLAB scores to make mutually agreeable language switches.

The DLPT is under recalibration, but, if the most-qualified recruits are failing at a 50-percent rate, are expectations too high? Is this specialty too demanding for enlisted personnel? Or could students who fail to reach required language proficiency be used in some restricted mode and continue to learn the language on the job?

Recent Air Force Actions

The Air Force has significantly changed the phase program, shortening each phase, and providing more privileges sooner (2AF, 2010). Further discussion is found in Chapter Ten.

Changes to the DLPT5 have resulted in an increase in the pass rate from 35 percent to 70–75 percent (for Arabic).

CHAPTER SIX

Network Intelligence Analysis, AFSC 1N4X1

Overview

A network intelligence analyst analyzes, disseminates, and exploits intelligence derived from target network communications (AFMAN 36-2108, 2004).

Network intelligence training requires 93 training days and is taught in the 316 TRS at Goodfellow AFB. The course pipeline, illustrated in Figure 6.1, shows one training location for technical training. The 93 technical training days translate to 136 calendar days, approximately five months. A network intelligence analyst costs $37,000 to produce.[1] Candidates for the network intelligence specialty are required to achieve a score of at least 62 on the general composite of the ASVAB test. The general composite combines evaluations of verbal expression and arithmetic reasoning.

Table 6.1 lists the attrition rates by course. Currently, network intelligence trainees take one course with 12 blocks. AETC and the intelligence CFM are reworking all intelligence pipeline courses such that the curriculum will focus on information relevant to the first job and then add training as needed later.

Figure 6.1
Network Intelligence Analysis, AFSC 1N4X1, Training Pipeline

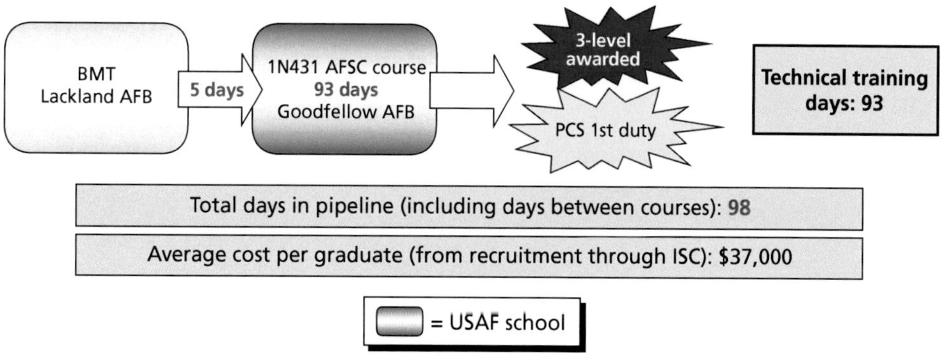

SOURCE: AETC/A3 (2007).

RAND *TR955-6.1*

---

[1] Factors are based on variable costs only and include the following: cost per graduate for training courses required for specific AFSCs at the basic skill level, acquisition costs (including the costs of recruiting, initial travel, and initial clothing issued), the cost of basic training at the Air Force Military Training Center (enlisted only), student pay and allowances, and time in transit and waiting for class to begin.

Table 6.1
Historical Attrition Rates for Network Intelligence Fundamentals Course (%)

Course	FY 2002	FY 2003	FY 2004	FY 2005	FY 2006	FY 2007	FY 2008
Network intelligence	18.2	19.7	17.8	16.6	12.7	14.4	11.2

SOURCE: TTMS.

Notable Factors

In the course of the interviews, the major issue that emerged was the length of the course. Other issues relating to job description and study time, although not unique to network intelligence, are also discussed in this section.

Length of Training

Several trainees, and one of the MTLs, remarked that the 17-week course was too long. Although trainees acknowledged that the course requires them to absorb a lot of information in a short amount of time, several felt that the overall pace of the training was too slow and covered material that would be better deferred to on-the-job training. To the extent that longer courses have higher attrition per se, adding even simple information to the course could increase attrition. The fact that the CFM was working to reduce the length of the course supports this contention.

The emphasis on memorization, and the fact that students disagreed on whether the pace was too fast or too slow, would suggest that self-pacing could be an effective strategy for this specialty.

Miscellaneous Concerns

There was general agreement that, as in operations intelligence, trainees do not know what they are getting into with network intelligence. Partly, this is unavoidable because the nature of the work is classified, so the career field description given to recruits is necessarily vague. But the mystique of a career in intelligence without knowing what it entails is bound to appeal to some recruits for the wrong reasons, which suggests that attrition might be reduced with prerequisites or a screening test of some sort. Trainees opined that having completed some college is advantageous for the study habits it instills. Others felt that the ability to memorize rapidly, puzzle-solving skills, and even creativity and intuition were most critical for success in network intelligence.

Additionally, several students and one instructor commented that the student rope program is not strictly voluntary, is stressful, and detracts from valuable study time.[2]

A related concern was the need to go to a secure facility to study classified material. Study time and the phase program also conflicted, with demands of the phase program occurring during times that could be used for studying.

[2] The rope program identifies students with special leadership responsibilities and who perform special duties. The green rope is the bay chief or element leader. The yellow rope is the flight leader, one step up from the green rope. The red rope is the student supervisor, the highest position an airman can attain in the squadron. There are other rope colors for chapel guide, drill team, and drum-and-bugle corps.

Finally, some students commented on a lack of sleep, some of it derived from the mandatory curfew (and associated bed-check) that essentially prevents recruits from going to bed early.[3]

Data Analysis

Forms 125A from network intelligence training were collected for the period between August 2007 and January 2009. Almost all attrition for the career field was due to academic deficiency, as Table 6.2 shows. Disciplinary problems were cited as contributing factors in nearly one-quarter of the cases, and family issues were noted in two cases.

The academic performance of network intelligence trainees is judged against the average class score of 93.8 percent. Students with a test-score average in the high 80s are still at risk of elimination if they have multiple exam or performance-check failures. Nearly half of the attrition (46 percent) occurred within 21 training days of entering the 93-day training pipeline (i.e., within the first five training blocks). Several airmen, including a student leader (black rope), were disciplined for repeatedly falling asleep in class or study hall, and one trainee blamed a test failure on lack of sleep. There was a single student eliminated for financial circumstances, which precluded him from obtaining the requisite security clearance. Another student's academic performance and military bearing seemed to deteriorate following a death in the family. He felt he was being targeted by MTLs and could not focus on his studies. Eventually, he was accused of deliberately failing the course and was discharged from the Air Force.

Analysis of the combined personnel/training database showed some correlation of training completion with mathematics knowledge, assembling objects, and age at enlistment.

The larger data analysis showed some potential improvement (10-percent explanation in variance) in selecting individuals for success in network intelligence (see Table F.8 in Appendix F). The two ASVAB subtest factors, mathematics knowledge and assembling objects, and the age-at-enlistment factor (older is better) were statistically significant in predicting success. But these showed only a small effect in predicting final grade, a 3-percent explanation of variation (see Table F.9 in Appendix F).

Table 6.2
Causes of Attrition for Network Intelligence Between August 2007 and January 2009 (n = 28)

	Attrition Rate (%)	
Cause	Listed Reason	Contributing Factor
Academic	93.1	
Discipline	3.4	24.1
Security issue	3.4	
Family issue	7.1	

[3] Mandatory curfew is common to all courses. It was mentioned in every AFSC interview but more often in intelligence-related courses.

Summary

The primary concern of students was the length of the training, approximately five months. Many students felt that the pace was slow, despite the tremendous amount of information to absorb, and that it would be better deferred to on-the-job training. Self-pacing could be an option in this course. Also, studying is difficult because of the classified nature of the work and the need to go to a controlled area to study classified material.

Additionally, students did not understand, before signing up, what the job involved. Understandably, many aspects of the job are classified, but students felt that a better description of the job was possible.

From a selection point of view, this career field, not surprisingly, sees greater success with individuals with technical competencies (mathematics knowledge and assembling objects). Age also correlates with success.

Recent Air Force Actions

The length of the course is now up to 110 days from the 93 days during the period of research. This lengthening contradicts the research conclusions and will need to be reevaluated in the future to see the effectiveness of the action. On the positive side, some of the material moved away from rote memorization of facts to application-based training (2AF, 2010). Additionally, changes in the phase program (discussed in Chapter Ten) could help improve motivation for a longer program.

CHAPTER SEVEN

Aerospace Ground Equipment, AFSC 2A6X2

Overview

AGE technicians are responsible for performing scheduled and unscheduled maintenance on machinery, such as hydraulic test stands, generators, and bomb lifts, in support of aircraft systems and subsystems (AFMAN 36-2108, 2004).

AGE, at 99 training days, is the longest training program in the 82 TRG at Sheppard AFB. The course pipeline, shown in Figure 7.1, utilizes one training location for technical training. The course is taught in the 361 TRS. The 99 technical training days translate to 147 calendar days, approximately five months. An AGE technician costs $40,000 to produce.[1] AGE requires ASVAB composite scores of 47 for mechanical and 28 for electrical. Recruiting seeks to fill 50 percent of the training GTEP enlistments; the remaining openings are classified during BMT.

Although attrition has been a problem in the past, recent actions have reduced attrition considerably. As shown in Table 7.1, the past three years have seen single-digit attrition levels, but washback rates have ranged from 17 to 30 percent.

Figure 7.1
Aerospace Ground Equipment, AFSC 2A6X2, Training Pipeline

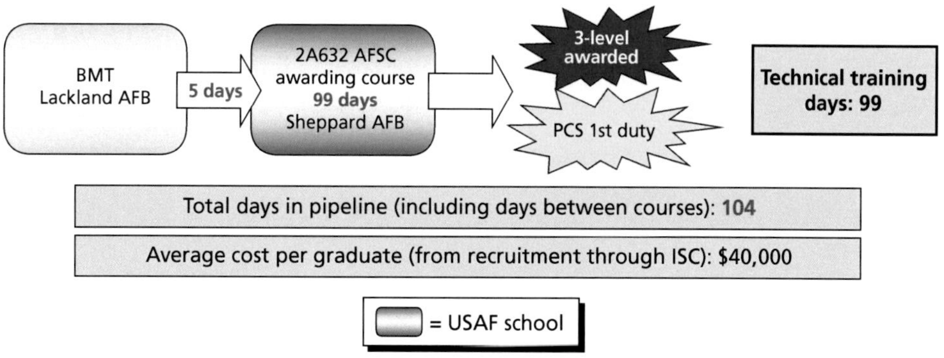

SOURCE: AETC/A3 (2007).

RAND TR955-7.1

[1] Factors are based on variable costs only and include the following: cost per graduate for training courses required for specific AFSCs at the basic skill level, acquisition costs (including the costs of recruiting, initial travel, and initial clothing issued), the cost of basic training at the Air Force Military Training Center (enlisted only), student pay and allowances, and time in transit and waiting for class to begin.

Table 7.1
Historical Rates for Aerospace Ground Equipment Training (%)

FY	Attrition	Washback
2001	9.1	38.0
2002	11.7	39.1
2003	14.0	18.7
2004	13.0	18.7
2005	6.3	19.1
2006	3.0	17.0
2007	4.4	30.3

SOURCE: 361 TRS data.

We see in Table 7.1 that washback rates have been high throughout the decade, ranging from approximately 20 to 40 percent. Attrition rates have been as high as 14 percent and as low as 3 percent. Although we might expect a correlation between high washback and low attrition rates in a given period, a test of the data showed no statistically significant relationship. We find that 36.4 percent of trainees who wash back more than once for academic reasons are eliminated from the program. Table 7.2 shows the breakdown of attrition rates by the number of academic washbacks.

Notable Factors

The two most noteworthy issues identified in the AGE career field were the high washback rate and the length of the course.

Washbacks Versus Attrition

Given the high washback and low attrition rates in the AGE course, the question arises regarding the true cost of washback versus the cost of attrition.

Table 7.2
Washback Outcomes for Aerospace Ground Equipment Training

Academic Washbacks	Graduate	Attrit	Total	Attrition Rate (%)
0	930	6	936	0.6
1	161	16	177	9.0
2	39	23	62	37.1
3 or more	29	16	45	35.6
Total	1,159	61	1,220	5.0

NOTE: In all cases of two or more washbacks, at least one of the washbacks was for academic reasons.

We examined the two costs using a mathematical model, which is described in Appendix G. Using this model to assess historical attrition and washback rates by block, we are able to construct Figure 7.2. The vertical axis is the increased costs over a 0-percent washback, 0-percent attrition case. The 361st has reduced its attrition rate to 4.4 percent, with a corresponding washback rate of 30 percent. Our results show that one student attriting costs the same as 2.2 students washing back. A student who attrits has a 0-percent chance of graduating, whereas washbacks graduate at an 80.6-percent rate. The 2.2:1 ratio implies that, if AGE cut its washback rate in half (down to 15 percent) by eliminating more academically deficient personnel, the attrition rate could increase to 10.9 percent without any increase in costs. Figure 7.2 illustrates this dynamic in that the cost increase over baseline with a 5-percent attrition rate and a 30-percent washback rate (about $460,000) is greater than the cost increase with a 10-percent attrition rate and a 15-percent washback rate (about $400,000).

We conclude that the AGE approach of allowing large numbers of washbacks is a cost-effective strategy for reducing attrition.

Length of Course

The majority of instructors interviewed felt that the course was too long, which contributes to attrition. One instructor mentioned 17 hours of database instruction, yet the students do not know how to use the database at graduation; he said they then go to the field and learn in less than one hour. Of course, it is not clear that the 17 hours do not contribute to a shorter learning time in the field.

One instructor felt that leadership did not want to cut course length because, "once cut, you can never get it back." Another instructor felt that the six-month course could be taught in six to eight weeks instead. A third instructor thought that the students did not need to learn

Figure 7.2
Increased Training Costs by Washback Rate and Attrition Rate

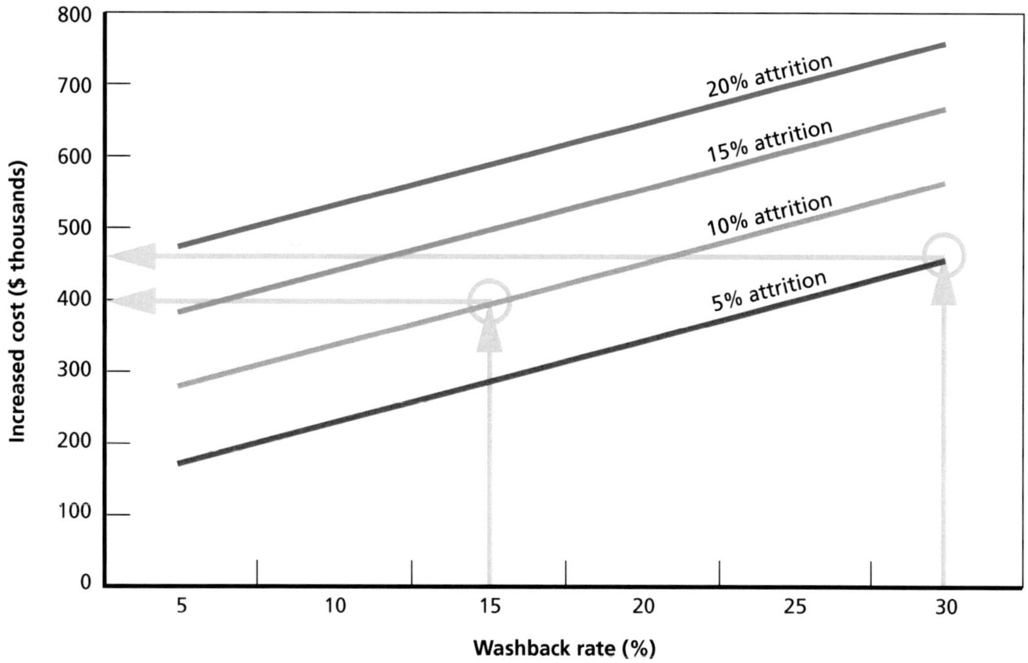

RAND *TR955-7.2*

every schematic for every system. Instead, this instructor proposed, they should just be taught how to read a schematic.

We did not ask the students their opinion regarding the length of the course because we did not believe that it would be very useful, yet one-third of the students also indicated that the course seemed too long.[2] They specifically mentioned a lot of dead time, unnecessary material, and learning systems that they will never use.

Although there is a relationship between length of course and attrition (see discussion in Chapter Ten), shortening the AGE course will not significantly reduce attrition because most of the attrition occurs up front (first few blocks), and the attrition is already low. Shortening the course would, however, lower costs.

Data Analysis

An examination of the 36 Forms 125A issued between October 2006 and November 2007 for the AGE specialty revealed that academic attrition was the leading cause of students being eliminated from the pipeline over this time frame. As shown in Table 7.3, 61 percent (22 out of 36) of the attritions were for academic deficiency, while nearly half that many (28 percent, or ten out of 36) were eliminated due to disciplinary issues. Four of the ten discipline cases involved abuse of drugs or alcohol.

The Forms 125A from Sheppard AFB were detailed enough to permit inference of contributing factors in the attrition. Disciplinary issues—in particular, violations of the phase program—were a contributing factor in 17 percent (six out of 36) of eliminations. Although family issues were not a primary reason for any attrition, they did appear as a contributing factor in 11 percent of the eliminations. We suspect that increased stress of family troubles contributes to the elimination of individuals who might not have been eliminated with lower levels of stress.

Typical of the AGE Forms 125A were reports of students failing multiple blocks or a single block multiple times, sometimes with disciplinary issues or family issues as contributing factors. One airman who was eliminated for poor academic performance had also received

Table 7.3
Causes of Attrition from Aerospace Ground Equipment Training Between October 2006 and November 2007 (n = 36)

	Attrition Rate (%)	
Cause	Listed Reason	Contributing Factor
Academic	61.6	
Discipline	27.8	16.7
Medical issue	11.1	2.8
Voluntary (SIE)	0.0	
Family issue	0.0	11.1

2 This answer was volunteered, not prompted; therefore, the proportion could actually be higher.

letters of reprimand for failing room inspections and possessing smokeless tobacco, as well as counseling for falling asleep in class. In another case, a trainee reported having trouble sleeping due to concern about relationship issues with his fiancée. Several airmen who were performing well academically were not only eliminated from AGE but separated from the Air Force for multiple phase violations. The airmen were reprimanded for violations, such as driving while on base, leaving the dormitory without permission, unauthorized spousal visitations, and failure to shave. Although a pattern of phase violations might reflect a cavalier attitude incompatible with military service, these violations might not constitute misconduct for airmen once they graduate from technical training. The Air Force should consider whether the elimination of students for deeds that are not per se impermissible is consistent with the original intent of the phase program.

The larger data analysis showed a very slight potential improvement (3-percent explanation in variance) in selecting individuals for success in AGE (see Table F.10 in Appendix F). Also, predicting final grade had almost no effect (see Table F.11 in Appendix F).

Exemplars

The 361 TRS has taken steps to examine and fix washback and attrition issues. A January 2008, 13-page memo highlighted the results of internal research on the cause of high washback. The research examined test questions, finding a high failure rate in certain questions and slight correlation with certain instructors. Additionally, they also examined the Flesch Reading Ease score of the student instructional reading material and found that the most-difficult material was at the beginning of the course, when students "are being introduced to brand new material in a brand new career field for the first time" (361 TRS Deputy of Operations, 2008).[3]

Summary

Washbacks are high, ranging from 17.1 percent to 30.3 percent over the past few years. Washbacks ultimately graduate at an 80-percent rate, so the policy of washing back versus attriting the student is a cost-effective strategy.

Our interviews revealed that the course might be longer than necessary. We would not expect much savings in attrition costs due to a shorter course, but we might see savings in reduced washbacks and would definitely see overall savings in basic course costs from eliminated training days and associated expenses.

Recent Air Force Actions

AGE was already in the process of changing training material and reviewing instructor performance during the time of the research. Recent changes to the phase program (discussed in Chapter Ten) should also improve attrition rates.

---

[3] Because the research findings had been implemented just before we visited the unit, we were not able to evaluate any of the effects.

CHAPTER EIGHT

Pararescue, AFSC 1T2X1

Overview

Pararescue personnel perform as the essential surface-to-air link in personnel recovery and materiel recovery by functioning as the rescue-and-recovery specialists on flying status as mission crew or as surface elements. They provide rapid-response capability and operate in the six geographic disciplines—mountain, desert, arctic, urban, jungle, and water—day or night, in friendly, denied, hostile, or sensitive areas. They also provide assistance in and perform SERE training. Last, they provide emergency trauma and field medical care, provide security, and move recovered personnel and materiel to safety or friendly control when recovery by aircraft is not possible (AFMAN 36-2108, 2004).

PJ training is 371 technical training days long and is managed by the 342 TRS at Lackland AFB. The course pipeline, shown in Figure 8.1, includes six training locations and nine courses for technical training. The 371 technical training days translate to approximately 24 months.[1] PJ training requires an ASVAB score of 44 on the general composite. Recruiting fills 100 percent of the openings through GTEP enlistments.[2] A PJ costs $250,000 to train.[3]

Admittance to the PJ program is contingent on passing the PAST, which is administered by the recruiter prior to job reservation and then again 30–45 days prior to entering active duty (Appendix D details the requirements of the PAST program).

Attrition rates by individual course are shown in Table 8.1.

Historically, the PJ career field has eliminated more than 70 percent of entries into the pipeline. However, most of the eliminations occur in the indoctrination course, the very first course in a multicourse pipeline (see entries by course for FY 2007 in Figure 8.2). The PJ strategy is a cost-effective approach for managing high attrition. If a career field is prone to high attrition, it costs less to experience the attrition up front in a course effectively functioning as a screener course.

---

1 The 24-month estimate includes the time students spend in between classes on "casual status."

2 The Air Force guarantees an additional 20 percent of PJ jobs to new recruits. The 100-percent requirement already includes projected losses for attrition, and this represents an additional planning factor.

3 Factors are based on variable costs only and include the following: cost per graduate for training courses required for specific AFSCs at the basic skill level, acquisition costs (including the costs of recruiting, initial travel, and initial clothing issued), the cost of basic training at the Air Force Military Training Center (enlisted only), student pay and allowances, and time in transit and waiting for class to begin.

Figure 8.1
Pararescue, AFSC 1T2X1, Training Pipeline

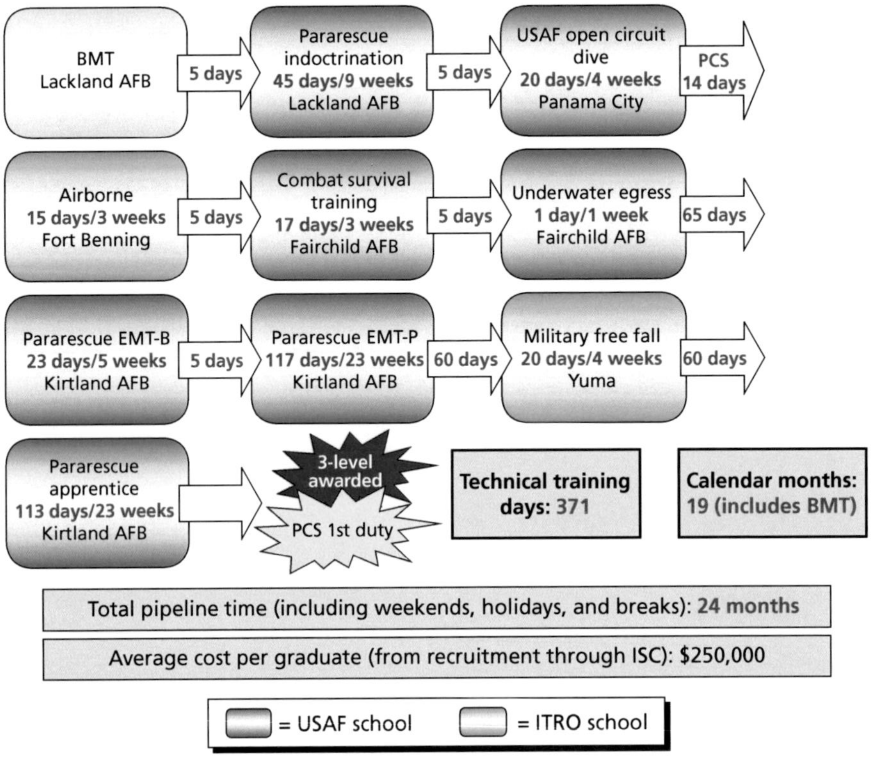

SOURCE: AETC/A3 (2007).
NOTE: In FY 2011, 2AF initiated a PJ development course prior to the PJ indoctrination course. EMT-B = basic emergency medical technician. EMT-P = paramedic EMT.
RAND *TR955-8.1*

Notable Factors

In our interviews and research of pararescue training issues, four unique issues stood out: the timing of the attrition, facilities, physical-fitness requirements, and issues of indoctrination.

Attrition Timing

From a cost perspective, the best place to attrit individuals is at the very beginning of the pipeline. PJ attrition overwhelmingly occurs during the first course, the pararescue indoctrination course. Reducing attrition during the indoctrination course, such that it is redistributed to other courses in the pipeline, is counterproductive and will only increase overall training costs. Therefore, there is no value in reducing the standards for success in indoctrination. On the other hand, reducing attrition in the indoctrination course by making trainees more successful is not counterproductive and will reduce the overall demand for students entering the pipeline. Consequently, consideration should be given to increasing the length of the indoc-

Table 8.1
Historical Attrition Rates for Individual Pararescue Training Courses (%)

Course and Version	FY 2000	FY 2001	FY 2002	FY 2003	FY 2004	FY 2005	FY 2006	FY 2007	FY 2008
Indoctrination 000	80.6	96.9							
Indoctrination 002		71.4	81.2	72.3	74.3	58.6	82.1	59.3	
Indoctrination 0P1A									66.9
Apprentice 002	8.0	0.0	5.9	0.0	15.2	8.8			
Apprentice 0P04A							0.0	1.8	
Apprentice 0P04B								0.0	4.0
Basic EMT 002				0.0	2.0	1.4	0.0		
Basic EMT 0B0A								2.0	2.7
Paramedic EMT 003				18.2	22.2	11.0	8.9		
Paramedic EMT 0P0A								2.0	6.1
Dive 0D0A							32.8	35.8	27.3
Dive 010A									14.7

SOURCE: TTMS data.

NOTE: These are not all of the courses in the PJ pipeline. EMT = emergency medical technician.

Figure 8.2
Entries by Pararescue Course

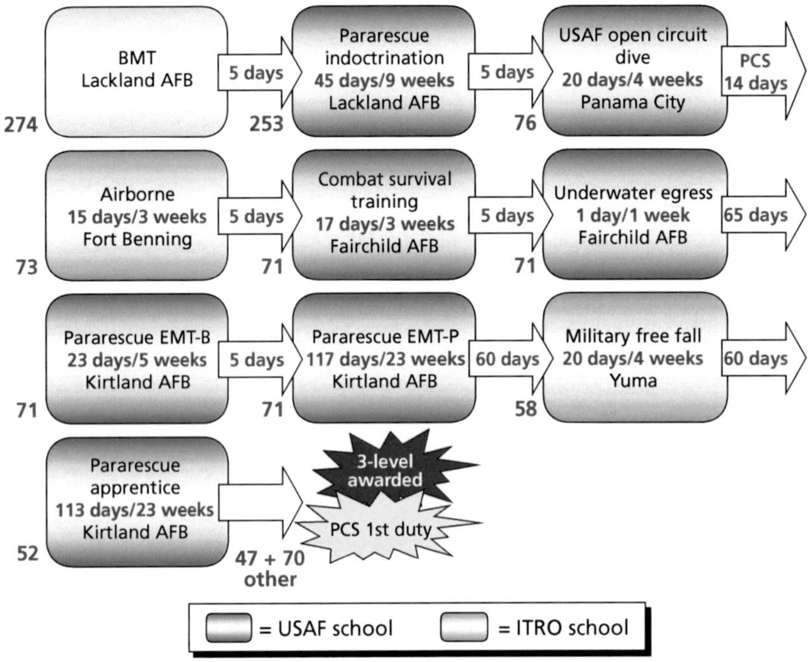

NOTE: Numbers to the lower left of each course (represented by a block) are the number of entries into the course. Likewise, the number prior to the next course would represent the number of successful completions of the previous course.

RAND *TR955-8.2*

trination course or to offering a separate course for trainees who are at the margin for physical qualification.[4]

Facilities

The facilities were crowded and in marginal condition. There is no pool dedicated to battlefield-airman training. The base pool is located some distance away from other course facilities and is not available for non–duty-time usage. Facilities issues were not directly linked to attrition or retention, although they would not be cited as helpful either.

Physical Fitness

The majority of PJ and CCT students interviewed felt that BMT hurt them physically. One felt that his run time had improved but that everything else worsened. In one discussion with five eliminated trainees, all five felt that BMT hurt their physical ability. Trainees specifically mentioned certain categories of physical training that were not conducted (chin-ups versus pull-ups) or were too easy (long-distance running, push-ups, and sit-ups). One trainee indicated that he tried one time to do extra push-ups and was reprimanded by the TI for being an "individual."

The Fit Flight concept appears to be an excellent idea with poor implementation. At the time of the interviews, Fit Flight trainees in BMT (PJ, CCT, and SERE) were given blue arm-

[4] Physical fitness is the primary determinant of success in the PJ indoctrination course.

bands and dispersed among multiple flights in two specific squadrons. Fit Flight trainees were interspersed with non–Fit Flight trainees. The interviews suggested that the TIs picked on them a little more but that the Fit Flight trainees did no additional physical activities.

A better Fit Flight concept would be to create entire Fit Flight flights. These flights could do more physical activities without worrying about the effect on non–Fit Flight trainees. The argument against this idea is that it creates "elite" airmen, whereas BMT is trying to equalize everyone. This is a subjective decision, but it would appear that training elite airmen to a higher level in BMT would reduce attrition in follow-on technical training. Another argument is that the Fit Flights would always win the commandant's award for best flight. Interviews with pararescue instructors indicated that the commandant's award was not important and that the Fit Flights could be excluded. We disagreed that the commandant's award would not be important to Fit Flight trainees and believe that another solution would be to handicap scores related to fitness for the Fit Flights.

Recruits are required to pass the PAST to qualify for the pararescue and combat control career fields. Baumgartner (2008) studied fitness levels among GTEP special-tactics trainees, including combat control, pararescue, survival specialist, and TACP specialties, and found that BMT has a positive effect on physical fitness for most GTEP trainees but a negative effect for some trainees in elite tactical career fields, such as PJ, CCT, SERE, and TACP. His three explanations for the relatively low PAST success rate among even special tactics trainees early in BMT are (1) recruiters sending unfit candidates into the pipeline, (2) trainees losing fitness while awaiting entry into the pipeline, and (3) failure of BMT instructors to teach and demand proper execution of physical training.

Baumgartner also demonstrated that performance on certain PAST events is strongly correlated with performance in the initial technical training course for all of the battlefield-airman schoolhouses (pararescue, combat control, and tactical air control). Specifically, out of a cohort of 800, only one-third of the 206 trainees who fell below the PAST standard on the 1.5-mile run graduated from the technical training course, and only 28 percent of the 131 trainees who fell below the PAST standard of six pull-ups graduated. Baumgartner estimated that, all else equal, if all trainees in the sample of 800 had met the PAST standards for the run and pull-up events at the beginning of their technical training course, the number of successful graduates (in the indoctrination course) would increase by 34 and 25 individuals, respectively. Based on the attrition rates over the remainder of the pipeline, 100-percent pass rates for the run and pull-up events would increase production by 20 and 16 additional training-program graduates, respectively.

With regard to PJ trainees in particular, Baumgartner collected results for a PAST on the first day of indoctrination for 277 PJs and tracked their performance in the course. Differences in mean PAST times or repetitions between graduates and nongraduates were highly significant for all events. None of the 98 graduates failed to meet the PAST standards for the 1.5-mile run, the push-ups, the sit-ups, or the 500-meter swim, and only one failed to meet the pull-up standard. By contrast, 10.6 percent of the 179 nongraduates failed the 1.5-mile run, and 22.3 percent failed the pull-up event (the pass/fail differences for other events were less dramatic). In terms of the likelihood of passing indoctrination, 38 percent of those who met the 1.5-mile run standard (and 0 percent who did not) graduated from the course, while 42 percent of those who met the pull-up standard (and 5 percent of those who did not) graduated from the course.

Baumgartner's works suggest that the primary way to decrease attrition in the indoctrination course is to improve physical fitness in BMT.

Indoctrination

In our interviews, some of the students mentioned a negative attitude among certain TIs (during BMT) toward the job of pararescue. There was no indication that this is widespread, but it was mentioned by multiple interviewees. Also, there are no pararescue (or CCT) personnel serving as TIs in BMT. The interviews suggested that having a battlefield-airman TI would have been beneficial for morale and might have prevented some of the day 1 losses. We would recommend that one or two battlefield-airman TIs serve in exclusive Fit Flights.

Data Analysis

Fifty-four trainees were eliminated from the pararescue indoctrination course at Lackland AFB between April 2007 and May 2008, and another nine were eliminated from the EMT and apprentice courses at Kirtland AFB. As Table 8.2 shows, voluntary disenrollment was overall the most common reason for attrition during this period, at 52.4 percent, followed by medical issues, many of which were injuries incurred during the rigorous physical training required for this career field. Indeed, most of those who self-eliminated were described as lacking the physical and mental fortitude to endure the intense training regimen of pararescue. Physical deficiencies and extreme aversion to water drills (hydrophobia) were thus the two most common contributing factors for indoctrination-course attrition. Voluntary attrition tends to occur significantly earlier in the training (mean = 7.5 days) than attrition for other reasons (mean = 27.5 days; $t = 3.4$, $p < 0.01$). Further along in the pipeline, when the physically unfit students had selected out, academic and discipline problems came to the fore. Of the eight discipline-based eliminations, six were in the later stages of the pipeline, and, of those, four were for incidents involving alcohol.

Several of the Forms 125A for the indoctrination course cited lack of physical and mental preparation, desire, or confidence. Two airmen who had self-eliminated due to their inability

Table 8.2
Causes of Attrition for Pararescue Between April 2007 and May 2008 (n = 63)

	Attrition Rate (%)	
Cause	**Listed Reason**	**Contributing Factor**
Voluntary (SIE)	52.4	
Medical issue	27.0	1.6
Discipline	12.7	1.6
Academic	4.8	
Physical	3.2	22.2
Family issue	0.0	
Hydrophobia		23.8

to withstand the water training were described as exhibiting "overt panic" and "uncontrollable anxiety" during drills in the pool; one was actually referred to a psychologist. In terms of academic attrition, two students failed the National Registry of Emergency Medical Technicians course, and one was eliminated after failing the USAF combat dive course three times.

The larger data analysis showed a small potential improvement (7-percent explanation in variance) in selecting individuals for success in pararescue (see Table F.12 in Appendix F). Three of the factors were physical in nature, which is not surprising. Additionally, the ASVAB mathematics knowledge composite was statistically significant, although it is not clear why.

Finding personnel factors that correlate with predicting a higher final grade had a small improvement (6-percent explanation of variance), but only mathematics knowledge was significant in predicting final grade (see Table F.13 in Appendix F).

Exemplars

The 342 TRS phase program at Lackland AFB was well managed and had a positive impact on retention. The PJ and CCT students undergo tremendous stress and physical conditioning. Their MTLs acted as mentors and a support system to the students. They enforced discipline but did not add stress to an incredibly demanding program.

Summary

Previous research shows that certain levels of physical fitness are directly correlated with success in the indoctrination course. At the time of research, the Air Force had just instituted the concept of Fit Flights. We believe that the concept of Fit Flights is good but that the execution was wrong. We recommend that Fit Flights be segregated in order to conduct physical training (PT) at a higher level.

Quantitative data support the current concept of using the PAST to assess a person's qualification for pararescue training, as physical factors were significant factors. The EQ-i might also provide additional noncognitive information for selection.

Recent Air Force Actions

The Air Force is activating a PJ development course to precede the PJ indoctrination course for students whose physical fitness is on the margin. Also, BMT Fit Flights include additional fitness training, swim sessions, and mentorship with CCT and PJ staff during BMT (2AF, 2010).

The Air Force is now administering the EQ-i for all battlefield-airman volunteers at the MEPS with the goal of using the results as a screening tool later (2AF, 2010).

CHAPTER NINE

Explosive Ordnance Disposal, AFSC 3E8X1

Overview

An individual working in the EOD career field is required to locate, identify, disarm, neutralize, recover, and dispose of hazardous explosives; conventional, chemical, biological, incendiary, and nuclear ordnance; and criminal or terrorist devices. The person also performs, supervises, and manages EOD operations (AFMAN 36-2108, 2004).

EOD training is 151 technical training days long and is managed by the 366 TRS at Sheppard AFB. The course pipeline, shown in Figure 9.1, includes two training locations and three courses for technical training. The primary course at Eglin is administered by the Navy. The 151 technical training days translate to 225 calendar days, approximately eight months. It costs $63,000 to produce a graduate.[1] EOD requires an ASVAB composite score of 64 for general (a relatively high score).[2] Recruiting fills 100 percent through GTEP enlistment.

Figure 9.1
Explosive Ordnance Disposal, AFSC 3E8X1, Training Pipeline

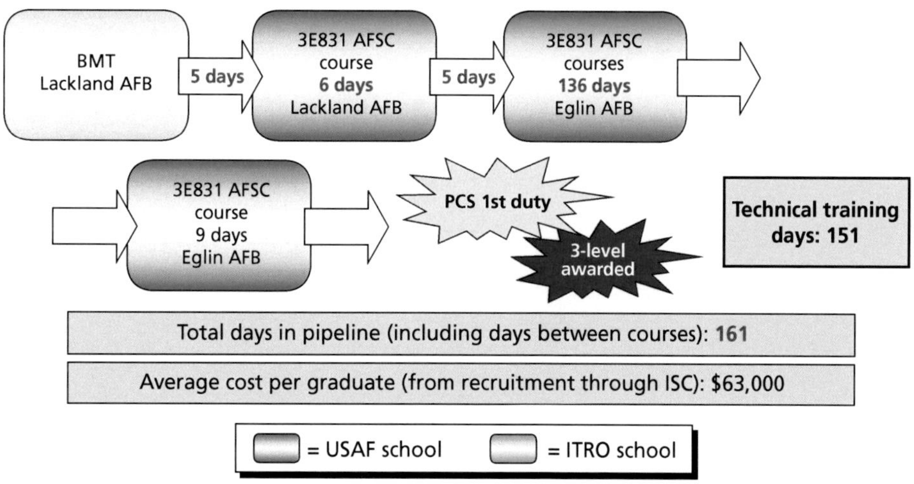

SOURCE: AETC/A3 (2007).

RAND *TR955-9.1*

---

[1] Factors are based on variable costs only and include the following: cost per graduate for training courses required for specific AFSCs at the basic skill level, acquisition costs (including the costs of recruiting, initial travel, and initial clothing issued), the cost of basic training at the Air Force Military Training Center (enlisted only), student pay and allowances, and time in transit and waiting for class to begin.

[2] Out of 69 AFSCs with ASVAB general score requirements, only 13 have similar or higher entry requirements.

The attrition data presented in Table 9.1 show that attrition has averaged 33 percent over the past eight years, with a low of 19 percent and a high of 46 percent.

Notable Factors

In our interviews and research on EOD training issues, some unique issues stood out: the input of students sent to the school, the maturity of the student, the need for a screener, poor facilities, and the phase program.

Production

In terms of EOD production, the problem is not only attrition but also placement. The Air Force has filled only 85 percent of its authorized seats over the past seven years and only 70 percent over the past four years (Figure 9.2). In the next section, we make an argument for using more PS and retrainee individuals, like the other services do. This could help fill the empty seats and, based on the analysis that follows, should reduce overall attrition.

Table 9.1
Historical Attrition Rates for Explosive Ordnance Disposal Training (%)

Course	FY 2000	FY 2001	FY 2002	FY 2003	FY 2004	FY 2005	FY 2006	FY 2007
EOD apprentice	36	19	30	24	36	40	46	33

SOURCE: 366 TRS Detachment 3 data.

Figure 9.2
Air Force Course-Entry Fill Rate

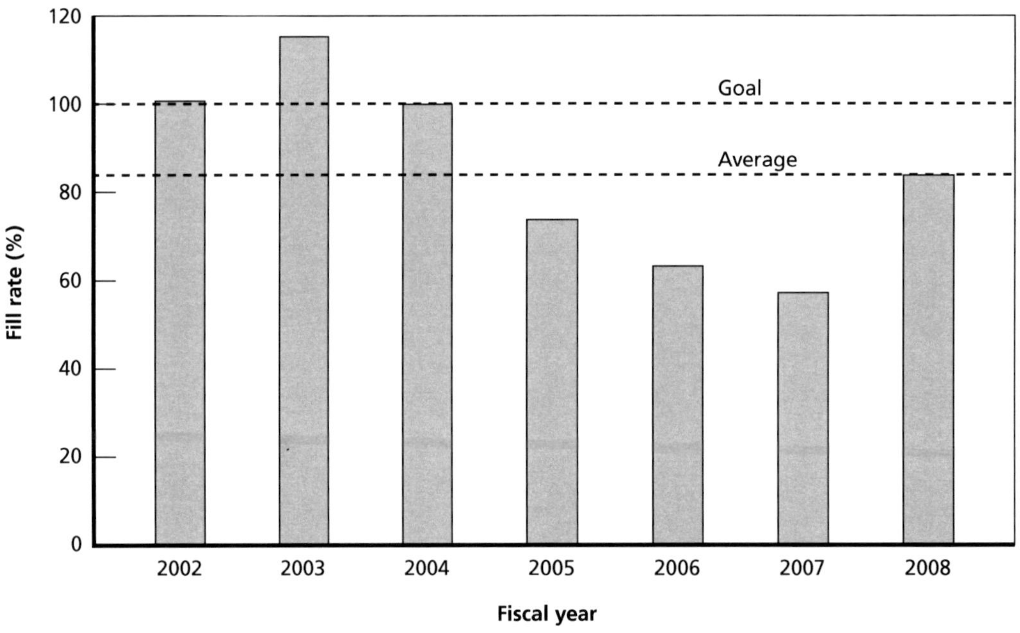

RAND *TR955-9.2*

Maturity of Input

There is a direct correlation between rank/age and attrition during EOD training. In his dissertation on stress and other risk factors in EOD failure, Bates (2002) found age to be a significant predictor of success. The author studied a cohort of 498 enlisted personnel who trained in EOD at Eglin AFB between August 1999 and October 2000. As seen in Table 9.2, Navy and Marine trainees were both significantly older and significantly more successful than USAF trainees. The performance discrepancy is all the more striking in light of his finding that USAF students had significantly higher scores in cognitive ability than Marine and Army students did. When age was the only factor in the model, the probability of successfully completing the training pipeline increased by 5 percentage points for every one-year increase in the age of the trainee. Pay grade, which is correlated with age and military experience, was an even stronger predictor of EOD success, with an odds ratio of 1.33.

Bundy and Sims (2007) evaluated the cognitive characteristics of successful military and public-safety bomb-disposal technicians using the Multiple Intelligences Developmental Assessment Scales (MIDAS™) survey instrument. Of the eight categories of intelligence, EOD specialists rated highest in intrapersonal and interpersonal intelligence—that is, their understanding of themselves and of other people. More than the other six intelligence categories—musical, kinesthetic, math/logic, spatial, linguistic, and naturalist—interpersonal and intrapersonal intelligence are commonly associated with maturity and wisdom. These results accord broadly with the work of McCrae et al. (1999), who studied personality differences as a function of age in five cultures. The authors found that scale scores for conscientiousness and agreeableness increased markedly after adolescence. This work could explain why OS recruits are better candidates for EOD work than NPS are.

Returning to data we analyzed, Table 9.3 shows that USAF attrition rates by grade are similar to those of the Army. The Navy and Marines have considerably lower attrition rates.

Table 9.2
Interservice Comparison of Explosive Ordnance Disposal Trainees, 1999–2000

Characteristic	Air Force (n = 192)	Army (n = 183)	Marines (n = 46)	Navy (n = 77)
Mean age in years	21.3 (4.2)	23.1 (4.0)	23.8 (2.8)	26.7 (4.5)
Mean grade, E1–E6	2.1 (1.4)	3.1 (1.5)	5.1 (0.25)	4.7 (1.0)
Mean cognitive ability (0–100)	80	66	64	75
Program-completion rate (%)	65	64	76	90

SOURCE: Bates (2002).

NOTE: Standard deviations are indicated in parentheses.

Table 9.3
Explosive Ordnance Disposal Attrition Rate by Grade and Service (%)

Grade	Air Force	Army	Marines	Navy
E1–E3	37	36	—	13
E4	25	24	4	14
E5	13	13	7	6

The Navy's lower rates could be due to a ten-week series of two courses taken prior to EOD training (Table 9.4). However, the Army also utilizes a ten-week preparatory course without achieving the same favorable rates later on. The Air Force has a six-day preliminary course, in which more than 50 percent of students are screened out, yet it still has had the highest attrition rate downstream in the EOD pipeline for six of the past eight years (Figure 9.3). Additionally, the USAF preliminary course has twice the attrition rate of the Army's, even though it is only one-tenth the length of the Army and Navy preparatory courses (see Table 9.4).

The USAF six-day preliminary course has slightly higher attrition (56 percent) than that of the Navy ten-week preparatory course and dive school (58 percent).[3]

Table 9.4
Service Prerequisites to Explosive Ordnance Disposal Entry

Service	Description
Marines	Rank (E4–E5) and age (21+) requirements
Navy	2-week EOD preparation course: 32% attrition 8-week dive school: 14% attrition
Army	10-week EOD Phase I course: 25% attrition
Air Force	6-day EOD preliminary course: 56% attrition

SOURCE: 366 TRS Detachment 3 data. The USAF data match TTMS data for 2006–2007.

Figure 9.3
Explosive Ordnance Disposal Attrition Rate by Service

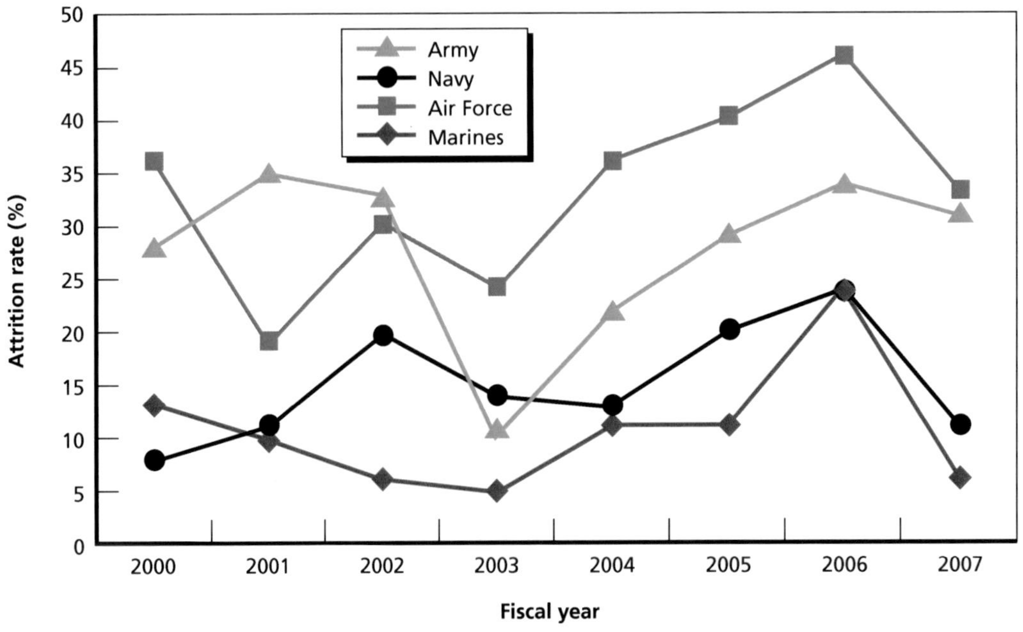

RAND *TR955-9.3*

[3] Navy attrition is calculated as (1.00 – 0.14) × (1.00 – 0.32), or 58 percent.

The Air Force sends the largest percentage of NPS students (those in grades E1–E3) of all the services, as seen in the FY 2007 snapshot (Figure 9.4). The Marines do not use any NPS students for EOD.

The Air Force's relatively heavy reliance on NPS recruits appears to be the primary factor driving attrition in the EOD course. In Figure 9.5, using FY 2007 attrition rates, we show that, if the Air Force decreased its NPS percentage from 77 percent (black arrows in Figure 9.5) to 55 percent (blue arrows in Figure 9.5), it could reduce its required input from 143 to 135 to produce the same 95 graduates. The resulting change in overall attrition is a reduction from FY 2007's 33 percent to 28 percent.

Increasing the number of PS trainees would reduce the demand for EOD recruits from BMT, and, since PS airmen are statistically more successful in the six-day preliminary course, input for that course could be reduced as well. In Table 9.5, although reducing the NPS percentage from 77 to 55 saved only eight entries (NPS and PS) into the Eglin course (143 to 135), it reduces the NPS accessions from 281 to 199 recruits, saving 82 recruits. It does require an additional 32 PS entries (33 to 65 entries overall; overall savings would be 50 [i.e., –82 + 32] less individuals sent to EOD to produce the same 95 graduates).

Interviews with students and instructors indicated that there are PS individuals who would like to enter EOD. We recommend that EOD personnel visit security forces and civil-engineering squadrons to brief career field opportunities on a regular basis.

Need for a Screener

In screening candidates for EOD, the Air Force uses the mechanical and general composites from the ASVAB, and the minimum scores for acceptance are relatively high when compared

Figure 9.4
Explosive Ordnance Disposal Course-Entry Distribution by Grade and Service, Fiscal Year 2007

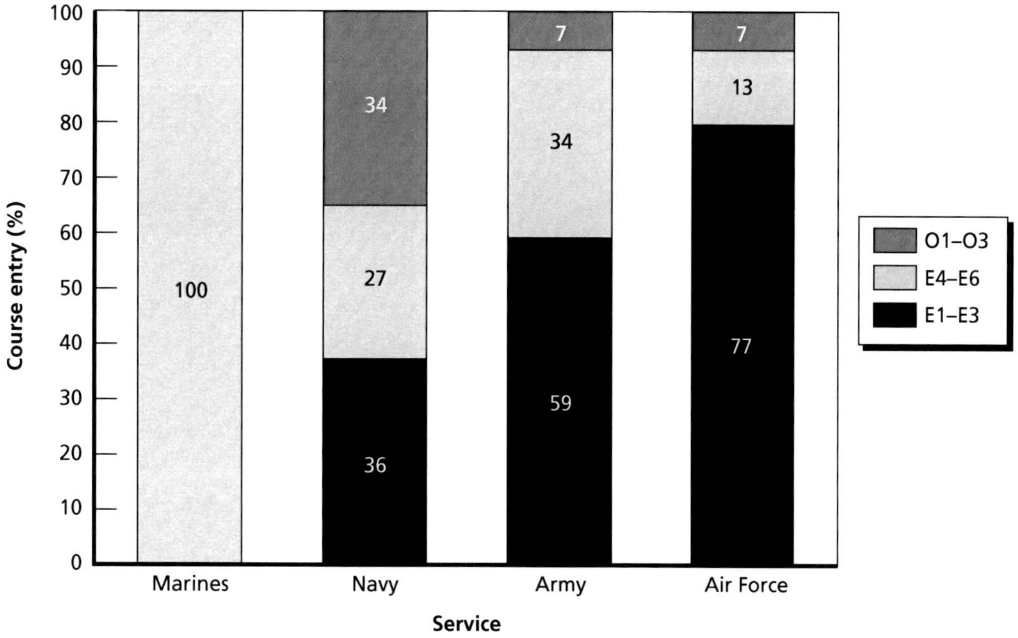

RAND *TR955-9.4*

Figure 9.5
Impact of Non–Prior-Service Percentage on Required Eglin Explosive Ordnance Disposal Course Input

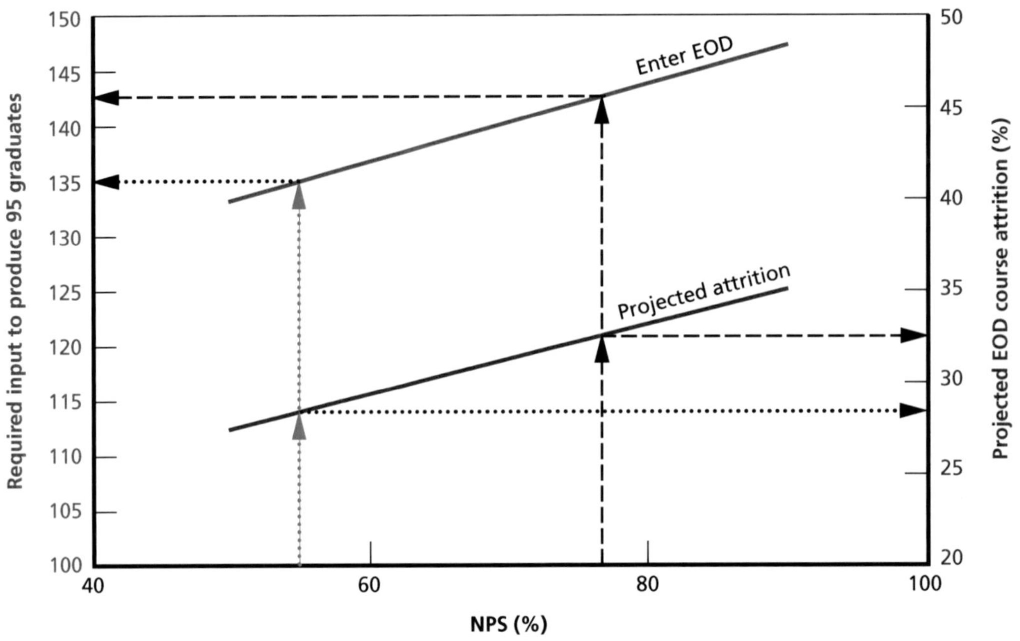

RAND *TR955-9.5*

Table 9.5
Effect of Non–Prior-Service Percentage on Accessions

NPS (share) (%)	Projected Attrition (%)	NPS Entries			PS Entries		
		BMT	Preliminary	Eglin	BMT	Preliminary	Eglin
77.0	32.7	281	260	117	n/a	33	26
65.0	30.4	235	218	98	n/a	51	41
60.0	29.4	217	201	90	n/a	58	46
55.0	28.5	199	184	83	n/a	65	52

NOTE: An NPS attrition rate of 55% was used for the preliminary course and 7.5% for BMT. For PS attrition, 20% was used for the preliminary course.

with requirements for most other career fields. However, evaluating other candidate characteristics could well be worthwhile. In a 1989 paper, Hogan and Hogan report on longitudinal research of Navy EOD diver students. In addition to cognitive measures from the ASVAB, the subjects were evaluated using a battery of tests that included three other classes of measures: personality, vocational interests, and physical performance. Subsequently, the subjects were rated pass/fail in a 42-week EOD training course, as well as success in diver training. They found that neither word knowledge nor arithmetic reasoning scores from the ASVAB were significantly related to success in EOD training.[4] The noncognitive items were measured with a 310-item self-reporting test from which, based on correlations with success in EOD training, nine were chosen. Six items were positively related to the criterion, and three were negatively related. The positive items were science ability, not being depressed, self-confidence, not being thrill seeking, leadership, and competitiveness. The negatively related items were structure, planfulness, and liking crowds.

These results suggest that well-constructed personality tests (such as the EQ-i) could be useful in identifying successful candidates for EOD training.

Facilities

The dormitories available for students in the EOD course were well below typical USAF standards, though we noted that all four services suffer equally. Building construction was ongoing during our visit and should improve some of the facilities. We saw no evidence that the condition of the facilities affected attrition, though students and instructors did raise the subject in the interviews.

Phase Program

The Air Force is the only service running a phase program in EOD. The Air Force has a higher percentage of NPS students than the other services, and that fact probably contributes to the decision to run such a program. Some of the EOD staff felt that the phase program contributes to high attrition. One of the staff shared an anecdotal case of a trainee being reclassified into a new career field and making length of training the sole criterion for his selection of reclassification choices (so that he would have the shortest possible phase program). We could not confirm any direct relationship of attrition to the length of the phase program. Further discussion of the phase program is found in Chapter Ten.

Data Analysis

There are two primary reasons for attrition from EOD: academic deficiency and voluntary disenrollment. Among the combined 189 Forms 125A generated from February 2006 to March 2008 at Eglin and Lackland AFBs, 77.2 percent listed academic deficiency as the reason, and 14.3 percent were for voluntary disenrollment. Voluntary disenrollments tended to occur earlier in the training pipeline (mean = 38.5 days) than academic attrition (mean = 53.3 days), but the difference was not significant. Forms 125A for voluntary disenrollments were sometimes annotated with comments to the effect that the EOD career field was not like the trainees had expected or been led to believe that it would be. Over this period, a small number of airmen were eliminated for disciplinary reasons (4.2 percent, eight of 189), family issues (3.7 percent, seven of 189), medical problems (3.7 percent, seven of 189), and security-clearance issues (3.1 percent, six of 189).

Figure 9.6 is a chart provided by the training detachment at Eglin AFB that supports the conclusions of the Forms 125A from Lackland AFB and Eglin AFB.

---

[4] Surprisingly, our investigation of data from a different period of time does show a statistically significant relationship between success and the word knowledge and arithmetic-reasoning composites. Still, that does not invalidate the conclusions regarding noncognitive testing.

The Eglin AFB Forms 125A were quite sparsely detailed, noting the reason for attrition and the recommended administrative action but without much description of the circumstances surrounding the attrition.[5]

The Forms 125A from the six-day screening course at Lackland AFB (n = 52) contained considerably more information, including students' comments. The Automated Explosive Ordnance Disposal Publication System progress check was clearly the biggest obstacle for EOD trainees at Lackland. Those who self-eliminated reported finding the job stressful and dangerous. Several airmen said that they were family oriented and that the danger and long deployments of EOD were incompatible with this orientation. Four of the Lackland AFB voluntary disenrollments complained that they had been either poorly informed or deliberately misinformed by recruiters about the nature of the EOD career field.

When combined with the Eglin voluntary disenrollments, 24 out of 189 (12.7 percent) felt ill informed about the job.

The larger data analysis showed a small improvement (6-percent explanation in variance) in selecting individuals for success in EOD (see Table F.14 in Appendix F). Five of the six significant factors were ASVAB subtest scores. The other factor is the ability to lift more than 100 pounds.

Predicting final grade had a significant improvement (20-percent explanation of the variance; see Table F.15 in Appendix F). The age of enlistment came in as an additional factor in predicting final grade, supporting the previous comments regarding the importance of maturity.

Figure 9.6
Distribution of Disenrollment Reasons

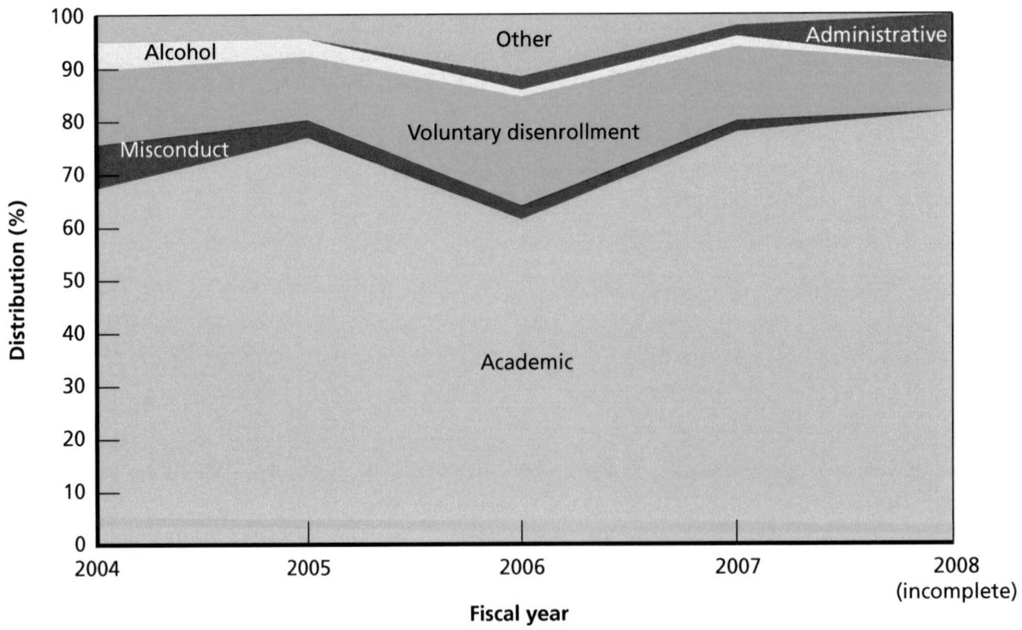

RAND *TR955-9.6*

[5] This is because we mistakenly did not ask for the Form 125A packages, only the Forms 125A themselves.

Exemplars

One day per month is set aside for routine appointments and errands that need to be accomplished during the duty day. The school is located at a training site, 45 minutes from the base. Additionally, all of the trainees are transported by bus, and there is little capability to commute to an appointment. This once-a-month training day provides some flexibility to the students for completing nonemergency and nonurgent appointments without falling behind.

Summary

USAF EOD students have the highest attrition rate of all four services at this joint school. This appears directly related to the proportion of NPS trainees. Increasing the ratio of PS students (in-service cross flows) should lower the overall attrition rate.

Previous research indicates that the use of noncognitive screening tools can reduce attrition. We recommended the implementation of a noncognitive test, potentially the EQ-i or alternative measure, prior to entry into the Air Force for EOD applicants.

Analysis of the quantitative data does show correlation between age and success in EOD for final grade.

Recent Air Force Actions

The previous six-day preliminary course was expanded to 20 days and moved to Sheppard AFB in the second quarter of FY 2011. Additionally, the test of a noncognitive screener, the EQ-i, started in the first quarter of FY 2011 (2AF, 2010).

CHAPTER TEN

Cross-Cutting Issues

In this chapter, we discuss several factors that could have a bearing on attrition for some or all of the career fields that we investigated. The factors examined here are

- recruiting experiences
- the phase program
- stress and training
- justice concerns
- operation of training bases
- consistency in documentation.

Recruiting Experiences

The recruiting experience can have an effect on schoolhouse attrition. Since all but one of the specialty fields that we investigated are in the GTEP, the inductees' initial skills training path is usually determined during the recruiting process. When the match between the trainee and the job is poor, the likelihood of the recruit failing to complete his or her training program is increased. Since these are demanding fields, it is important that the recruits understand what they are getting themselves into. In addition to attempting to steer people into fields for which they are qualified, recruiters have an obligation to provide information to help individuals make judgments about their suitability and taste for various specialties open to them.

Lack of information from recruiters about prospective USAF jobs was a complaint from 20 percent of recruits. About one-third of the students in PJ, CCT, EOD, and AGE claimed that their recruiters knew nothing about the respective career fields. More than half of the recruits claimed to have obtained information on their own initiative from such sources as friends and the Internet.

Many trainees stated that they were not given adequate information about their career field to sufficiently appreciate whether it was a good fit for them. In the CCT, EOD, and PJ courses, students commented that more information about the training would have helped them to mentally prepare themselves for the demands of training. Some also indicated that they were not given enough information about other potential jobs for which they might be better suited. A few indicated that they were given incorrect information about opportunities to switch to another career field or separate from the Air Force if they did not like the job or were eliminated from training. Some suggested that it seemed like their recruiters were just trying to meet job quotas. A few said that recruiters encouraged them to try to "game the system" by taking an available job and switching to the job they really wanted later.

Occasionally, instructors offered opinions about the recruiting process. An EOD instructor related that, in his view, recruiters attempting to fill quotas were not completely truthful. ATC instructors asserted that better screening by recruiters would be helpful; success in that field is known to depend on special aptitudes or personality types.

We also received many positive recollections. Numerous examples were mentioned in the interviews of recruiters who made extra efforts to properly educate potential recruits concerning the occupations under consideration. The best examples occurred in cases in which a recruiter knew someone in the career field and connected the recruit with the person. A recruiter takes a chance engaging a nonrecruiter. The individual might not be the best reflection of the Air Force or the career field, and it is possible to turn the recruit off from a career in the Air Force. Potentially, the recruiting service could engage a list of nonrecruiters from multiple career fields, minimally trained to recruit and to give useful information on the job and training.

The Phase Program

While in their technical training programs, trainees are also under restrictions imposed by the phase program. In brief, the program's goal is to transition a trainee from BMT graduate to a motivated, self-disciplined, and fit airman by progression through phases of greater responsibility and freedoms (AETCI 36-2216, 2004). At the time of the research, airmen spent at least 185 days (usually all of their training) in the phase program before becoming eligible for "phase grad" (14 days in phase I, 21 days in phase II, and 150 days in phase III).

Although we judge that the phase program is an important tool in the training of today's airmen, it is possible that the program might affect individuals' attitudes and, ultimately, attrition. If that is the case, modifications to the phase program could have benefits even beyond demonstrable improvements in attrition rates. In this section, we indicate noteworthy comments we received from trainees and instructors.

Student Comments

Students' attitudes toward the phase program were related to the length of time they had been in training and, in some cases, of their respective programs.[1] In general, students with less than four weeks of technical training were positive or at least tepid in their assessments of the program. Most of these trainees were in phase II. They understood the purpose of the phase program and felt that it was worthwhile. Students who had been in training longer than four to six weeks, however, were overwhelmingly negative (more than three out of four) toward the program. While agreeing that the phase program was of value in the initial period of training, they were quite emphatic in denying the continual need for it beyond six weeks. All of the training pipelines evaluated in this report were six months or more in length. A

[1] An exception to this generalization occurred in the case of students at Lackland AFB assigned to the 342 TRS, who were positive in their assessment of the phase program even after having been in the program for six weeks. The 342 TRS MTL program, at the time of the report, avoided adding extra stress through the phase program to the PJ, CCT, and EOD trainees during their initial course due to the tremendous amount of stress already in the training course.

source of frustration expressed by some students was that friends in career fields with shorter training programs were already at their new bases being treated as "adults" while they were still being treated like "children." A sampling of comments by students who had been in the phase program longer than six weeks included the following: It is a joke; standards are not the same; it creates hassles in dorm life that affect morale and stress; it has dumb rules; it allows double jeopardy; there is no time to study (a common complaint); and it causes people to be stressed out.

Ultimately, some trainees decided that the best strategy was to never come out of their rooms. One medical trainee volunteered for extra clinical phase II training in order to stay out of the line of sight of MTLs and out of trouble.[2]

Instructor Comments

Instructors were much less inclined than students (approximately 10 to 20 percent) to express negative opinions of the phase program. Specific comments included the following: Students do not have time to study (three instructors); it is too rigid for six months; it is too structured; it has too many rules; the rules are inconsistently and unfairly applied; it is a burden that, after six to eight months, causes burnout (MTL comment); it gets old; and there is room for improvement.

Comments with Respect to Military Training Leaders

Somewhat fewer than half of the trainees offered negative comments regarding MTLs, including the following: They take their jobs too seriously; they make too many rules; they unfairly target certain individuals; they are on a power trip; they are "over the top"; they publicly embarrass trainees; a high-and-tight haircut was considered too long; and it is too easy to phase back.

There were also complaints about student leaders or "ropes." Most of the gripes expressed lack of respect arising from perceived abuse of authority in playing favorites and helping friends.

Possible Modification of the Phase Program

Two prominent issues regarding the phase program are its length, particularly for career fields with long training programs, and that trainees expressed anxiety about being put back to a prior stage. An insightful comment was, "[I]f you didn't have to worry about phasing back, it would make it much easier."

Due to the amount of time airmen currently spend in phase III (essentially until graduation) and the feedback we received from both trainees and staff, an idea for consideration is introducing a phase IIIB, which students would enter after a certain amount of time in phase IIIA. These students would still be in phase III but could be phased back only for egregious offenses, such as those requiring an Article 15.

Another proposal would be to simply end the phase program after a fixed amount of time (we suggest six to eight weeks) for all career fields.

---

[2] In this case, *clinical phase II training* refers to nondidactic medical training in which medical trainees assist at the hospital. (*Didactic* in this context means instruction, such as lectures, as opposed to hands-on practical exercise.)

Stress and Training

Many of the comments we received, particularly with respect to the phase program, reflected feelings of being stressed. This led us to briefly examine literature on the relationship between stress and training.

Demands and Expectations of Military Service

Military service is demanding, and service members are exposed to physiological (long days), cognitive (technical knowledge), and psychological (low control) stressors that can affect performance and health. Many USAF occupations (e.g., ATC) do not require direct combat but still require preparation for enduring extreme stressors and thus require training for high-level performance under such conditions (Staal et al., 2008). Training for military service is necessarily rigorous because service personnel must be able to perform under stress in order to survive extreme and dangerous conditions. In addition to strengthening survival skills, basic training is designed to identify recruits who are unlikely to be resilient under stressful conditions. Attrition from basic and technical training programs is high, with more than one-third of recruits leaving military service before the end of the first term of enlistment (Knapik et al., 2004; Buddin, 2005). Special training for high-stress occupations might be especially challenging. Certain occupations within the military, such as EOD, are particularly intense and subject to high-stakes failure; only 40–50 percent of trainees graduate from the six-day selection and assessment course.

Performance Under Stress

Stressors include both physical stress resulting from heavy workload demand (e.g., long hours and length of training) and psychological stress, which can stem from myriad challenges, including cognitive demand required to learn new skills, uncertainty and insufficient information or unknown expectations about the future, and perceptions of injustice. Stress can be expected to have a significant impact on performance that can be even greater in the current high-technology military environment, which is characterized by complexity, intensity, and danger (Driskell and Salas, 1996). However, some factors serve as moderators to stress, including military leadership and unit cohesiveness (Belenky, Noy, and Solomon, 1987). These moderating effects could also apply to the technical training context.

In the domain of stress variables, the problems or "hassles" of daily existence have proven to be better predictors of overall psychological distress than major life events have (Lazarus and Folkman, 1984). Affective distress has been associated with impaired job performance. Police recruits with elevated indices of psychosocial distress tended to receive less favorable probationary job-performance ratings (Cortina et al., 1992). Research indicates that social support is key to mitigating the effects of high perceived workloads (Kirmeyer and Dougherty, 1988) and that supervisors are ideally situated to act as buffers because they can provide the necessary support, information, and esteem (Greller, Parsons, and Mitchell, 1992). At its most severe, affective distress leads to depression, associated with impairment of memory and concentration (Burt, Zembar, and Niederehe, 1995). In the context of airmen in training, if the cadre and the MTLs are not providing this buffer or, as might be the case under the phase program, are actually contributing to daily hassles, they could be adversely affecting the performance of some trainees.

Fatigue—or, more colloquially, "burnout"—and sleep deprivation are two other stress variables that can adversely affect performance. Sleep deprivation has been shown to impair task performance, especially for tasks of greater duration, difficulty, and complexity (Tilley and Brown, 1992). Maslach and Pines (1977) found that burnout in mental-health workers was characterized by a decreased ability to cope with emotional demands. Bates (2002) points out that long training days and difficult tests can contribute to fatigue. This could be relevant for the AGE trainees, who continue to be eliminated for academic reasons even in the latter stages of the pipeline, when the material is supposedly easier.

A final category, situational stress variables, includes cognitive and somatic anxiety. Cognitive anxiety is concern about poor performance and its social consequences, while somatic anxiety represents the awareness of physiological symptoms of stress. The interaction of these two variables is thought to have a direct bearing on performance (Hardy, 1996).

Stress in Trainees

Research on the impact that stress can have on performance among EOD trainees found that the relationships between enduring stress from inattention and situational cognitive anxiety alone were not associated with performance decrements. However, the additive effect of having personal or family problems was linked to poorer performance (Bates, 2002).

The literature on training under pressure has shown that high levels of stress can have deleterious effects on cognitive performance and suggests that training that balances learning with hands-on preparation might be most effective (Keinan, Friedland, and Sarig-Naor, 1990). Moreover, individuals learn better under minimal stress, and principles of learning environments suggest that phased training, in which initial training is provided under minimal stress conditions to maximize cognitive processing followed by more-stressful field training (Staal et al., 2008), might be more useful than pairing high-intensity physical demand with cognitive demand.

Implications for Resilience

Given the stresses of military training and service, programs that emphasize strategies to prevent adverse reactions to stress are key. Like the bourgeoning field of programs designed to promote resilience in service personnel, programs for training future airmen could benefit from devoting a portion of the curricula to strategies for successful completion of technical training programs, particularly those of longer duration or intensity. It is important to incorporate components of resilience that emphasize both physical and mental fitness. This is because even those who pass mental-fitness screenings are still subject to performance decrements and adverse psychological reactions later in their careers. In fact, training that focuses on strategies to promote resilience for particular types of stressors faced by service members in high-stress occupations might lower stress reactions and, ultimately, could reduce attrition. Such teachable resilience-enhancing efforts might include equipping instructors with the skills to create resilience in trainees (Campbell, Campbell, and Ness, 2008).

Resilience is critical not only to warfighting but also for successfully completing technical training programs. Many of these programs require rigorous physical and psychological conditioning. To maximize the value of such training, the military might consider incorporating lessons from the literature on learning environments by using phased training described earlier (initial training under minimal stress followed by more-stressful field training). Finally, an examination of how incorporating moderators to stress, such as leadership and cohesiveness, into technical training curricula could potentially promote resilience and ultimately reduce attrition.

Justice Concerns

Perceptions of justice have been shown to affect attrition in a variety of jobs (Finn and Lee, 1972; Schaubroeck, May, and Brown, 1994; Simons and Roberson, 2003). Some of the observations in the preceding sections on recruiting and the phase program can be interpreted as perceived injustice, and it is possible that trainees' perceptions of unfairness in the recruiting and training processes could account for some of the attrition in these career fields. These issues are explained in greater detail in Appendix D.

Trainees could easily regard the incidents of incomplete, misleading, or inadequate information recounted in the discussion of recruiting as examples of unfairness.

It was noted in the section on the phase program that trainees are subjected to its restrictions until the completion of their technical training. Because there are wide differences in the lengths of training programs among the various career fields, so too there are large differences in the length of time that new members of the Air Force are in the phase program. This situation can easily be taken as being unfair. Moreover, adding to the perception of injustice is the fact that many of the most-selective and complex training programs have the longest training times. As a result, personnel who perhaps are deserving of greater latitude due to the complexity of their career fields are given less latitude because of the extended phase program.

A third kind of perceived injustice was raised by EOD trainees. This very short and intense course is designed for screening; the actual technical course (run by the Navy at Eglin AFB) follows. The goal is to eliminate, as early as possible, individuals who are not likely to be successful in the career field. Students are required to memorize detailed technical material quickly without the opportunity to study after the end of training day. Moreover, students are required to execute practical exercises with no mistakes permitted. Although some students regard these demanding conditions as being unfair, the instructors assert that these are circumstances under which EOD professionals routinely operate. Students' sense of injustice could be mitigated by providing more-complete information about the demands of the job and the reason for the rigorous testing. Better advice and enhanced screening at the time of recruiting could serve to lessen the perception of unfairness at being given a job only to have it taken away in the first week of training.

Training-Base Purpose

We observed that, at some training bases, support functions maintain hours more consistent with a non–training-base operating environment, i.e., they close at about the same time the training day ends. In a nontraining environment at a nontraining base, airmen often have the freedom to take time off from work to run errands, but, at a training base, the airmen do not have the same freedom to run errands during the training day. Consequently, in the absence of extended hours in the support facilities, trips to the post office, military-clothing alterations, and military-personnel flight activities have to fit into the lunch hour, provided that the office is not closed during lunch. Airmen can be excused from training for a doctor's appointment or to pick up a prescription, but that does not serve the purpose of training. Training-base support services' hours ought to fit the needs of the primary population on the base: the trainees. This is the case at Lackland AFB, although probably because of BMT rather than technical training.

Length of Training Day

In the eight-hour training day, medical appointments occur during training. In the interviews, some instructors reported trying to adjust the instruction to account for a missing student, but this is clearly not feasible if multiple students have appointments spread throughout the training day. As a result, other instructors felt that it was better not to slow down than to risk getting behind.

Some courses have used the eight-hour training day for positive benefit. In the ATC radar course, instructors indicated that reducing the training day to six hours would have a negative impact because the extra two hours are used in simulators to reinforce classroom lessons.

A typical training day in these career fields often consists of an eight-hour day that is dominated by didactic instruction. Incorporating more-practical experience into the training day might be an effective strategy to increase graduation, performance, and retention in a career field. A switch from an eight-hour lecture day to six hours of lecture with two hours of hands-on training might also improve training outcomes. Based on research in nonmilitary settings, when newly learned competencies are transferred to the practical work environment, there is greater likelihood of increases in job performance (Velada et al., 2007).

An alternative to adjusting the mix of lecture and hands-on training is to spread training to include more reflection and absorption. Based on education research in elementary and high schools, it is unclear whether extending the school day has positive effects on educational outcomes (Pardini, 2001). Extensions of school days or school years also incurred greater costs. If there is a connection between absolute time spent in the classroom and achievement, the link is most likely in the amount of instruction that is engaging. Quality matters more than quantity (Pardini, 2001). In the USAF setting, this translates not to more time in didactic classroom settings but to increases in time for hands-on experiences designed to transform knowledge into work-related skills.

There is a prevailing sense that the current generation of recruits is very different from previous cohorts. The millennial generation is characterized by technological savvy. Previous generations have not been exposed to and raised in environments that are so infused with technology and virtual connectedness. Marc Prensky, an expert in the field of education and training of the millennial generation, argues that this generation processes information differently from the way any previous generation did. Prensky writes, "Our students are no longer the people our educational system was designed to teach" (2001a, p. 1). He also asserts that the individuals in this generation "think differently from the rest of us" and that "we now have a new generation with a very different blend of cognitive skills than its predecessors" (Prensky, 2001b, p. 5).

Others believe that the millennial generation "can deal with lots of information but prefers it packaged in short, focused segments. They want to know why but expect quick and direct answers" (Ricigliano, 1999, p. 124). To attract and retain the best possible individuals for the Air Force, it might therefore be necessary to adapt the ways in which new recruits are trained. In a recent white paper, the Air Force itself acknowledged the need to incorporate more technology in its training, to remain competitive in the labor market (AETC, 2008).

The notion that an incoming generation is radically different from previous ones and thus requires training and management of incoming staff is not new. When generation X was emerging into the workforce, there were concerns that the perceived apathy and laziness of the cohort would have negative consequences. Whether or not there were major problems with the integration of generation X is debatable. Today, to attract and retain members of the millennial generation, the infusion of technology into everyone's daily lives requires that the gap between "digital natives" and "digital immigrants" be minimized.

Although we cannot tie characteristics of the millennial generation to a best training-day length, the fact that high schools, vocational schools, community colleges, and universities do not use eight-hour training or instructional days suggests that the Air Force should reevaluate its training-day length.

Documentation

The usefulness of the Forms 125A as a source of data was diminished by variations across courses and training groups in the information they contained. MAGE scores and education level were usually noted, but not always. Sometimes, the justification for recommendations was articulated in detail; other times, the form merely noted that the individual had failed or self-eliminated without any further explanation. GTEP or prior technical school enlistee status is noted in some Forms 125A and not others. The inclusion of hours of special individual assistance in the Forms 125A varied by base and career field. These inconsistencies are mitigated by the fact that missing information can probably be obtained from other data sources, but, if the Air Force is reclassifying individuals without consulting this information, it could be setting airmen up for more failure.

Basic Military Training

The scope of research did not include BMT. We did interview staff and leadership at BMT as part of understanding the entire training pipeline. Our focus-group interview provided many insights and raised questions for future research.

Exemplars

Lackland AFB is uniquely structured to support new recruits. The base revolves around the initial recruit. Facilities that would not be open for extended hours on a normal base—e.g., base exchange, military clothing, alterations—are accessible late in the evening and on weekends.

At the end of graduation, there is a special ceremony, run by the enlisted corps, in which students are recognized as airmen and no longer trainees. On the negative side, they will soon again be "airmen in training" and not fully independent airmen (as the ceremony would have them believe) until they graduate from technical training.

Medical Holds

A significant number of students become injured during BMT. Two-thirds of the students placed in medical hold attrit out of the Air Force. Anecdotal comments suggested that the Army had a similar problem until it invested in the personnel to keep the recovering recruits motivated and on task to becoming soldiers.

Additionally, for efficiency, all BMT holds (medical, discipline, and others) are grouped together. The saying "one bad apple ruins the others" is especially true in this environment.

Classification

In the course of our investigation, we examined the computer program used during BMT to classify non-GTEP airmen in their career fields. We found that the program is very sensitive to changes in four weighting variables.[3] Additionally, little has been done to ascertain the effectiveness of the current classification program. The only metric currently produced is the percentage of recruits getting one of their five selections. This might be the best metric—a happy recruit would seem more likely to be a successful recruit—provided that the recruit has proper information about the career fields selected. This is an area for further investigation.[4]

Health Concerns

Anecdotally, there were cases of lack of sleep during BMT. This could be causal in health issues as well. When airmen arrive at their training base, a significant percentage are sent to the base clinic to recuperate. There has been some research into this area, but it bears further research.

Summary

In each of the nine career fields, 20 percent or more of the trainees specifically mentioned getting poor information from their recruiters. We recommend a review of the recruiting information sources. A majority of trainees agreed that more and better information from the recruiter would have been helpful, especially in terms of what the training would be like, allowing them to mentally prepare themselves for the demands of the training.

The student interviews revealed grievances with the phase program. In addition, the system is perceived as unfair. We believe that the phase program is good overall and should not be discontinued. But, although the phase program probably does not directly cause attrition, it is a key factor in adding stress to the recruit's daily life that, when combined with outside stressors, can cause attrition. We recommend multiple changes to phase III. At a minimum, at some point, it should not be possible to demote an individual to phase I, barring an Article 15 or some serious infraction. Additionally, we recommend making the phase program equal length for all career fields and suggest six to eight weeks as a common goal.

We recommend that training bases consider adjusting the work schedules of clinics, administrative services, postal services, and providers of clothing alterations to serve the trainee

3 These are (1) preference, up to 5 AFSCs (40 percent); (2) length of time to next class opening and the priority of the AFSC (40 percent); (3) desirable classes, as each AFSC has a set of desirable high school classes (10 percent); and (4) minimum ASVAB minus true ASVAB times 10 (10 percent).

4 During the course of the research, AETC SAS was researching the classification process and computer code used to classify individuals.

Although we have no specific recommendation about the eight-hour training day, some of the instructors interviewed suggested that information covered in eight hours could just as easily be taught in six. Reducing the training day to six hours would provide the trainees with time to run errands, make clinic appointments without missing instruction, and receive personalize instruction when facing difficulty with the material.

It became apparent during the course of the research that USAF As, which are used to record the reason for attrition, were not completed consistently.

We recommend further research into medical holds. We also recommend further research into job classification.

Recent Air Force Actions

The Air Force has made major changes to the phase program, reducing the time in each phase, increasing some privileges, eliminating "phase back," and reemphasizing the role of MTLs as mentors (2AF, 2010).

CHAPTER ELEVEN

Final Thoughts

In this research, we found that significant savings can be realized through reductions in training attrition and washbacks. Over the two-year course of the research period, we briefed numerous senior decisionmakers on our findings. Consequently, many of our recommendations have been or are in the process of being implemented. Those and other recent actions are documented in the preceding chapters.

The research showed that data investigation alone is not sufficient to understand the reasons for high attrition. The interviews and focus groups provided a wealth of information that was not apparent from an analysis of the personnel and training databases. It is more difficult for senior USAF personnel to extract the same information that the research team was able to glean. Because our research protocol afforded confidentiality, it allowed instructors and students to freely share their concerns. We were careful to validate individuals' comments. We noted students with a gripe or bad attitude, as well as students performing well yet having significant concerns about the program. We discussed the findings in depth before reporting conclusions. Given the value of the information, this type of research ought to be routinely considered as a tool for evaluating the health of training programs.

We also learned that, although there were concerns common across career fields, every career field we studied had unique issues. This underscores the need for caution in applying findings in one career field to the circumstances in other career fields, even if they are ostensibly similar.

Finally, one of the reviewers suggested a comparison of each service's practices for these unique and demanding AFSCs. Although we were able to do that for EOD training, as it is conducted by the Navy for all services, it would be a good idea for future research.

APPENDIX A

Interview Protocol

In this appendix are the questions we used in the interviews with instructors and with students. Depending on the answers and the time, we skipped questions or added a clarifying question during the interview.

Instructor Interview Questions

- Background
  - What is your background (military career, prior military, civilian service)?
  - When did you receive your training?
  - How have things changed or not changed?
  - Is the course harder or easier than when you attended?
- Students
  - What do the students like about the course?
  - What aspects do they find difficult?
- Attrition
  - What blocks cause the most attrition problems?
  - What are all the reasons why students attrit?
  - Describe some of the individuals who didn't complete the course and why?
- Students' motivation
  - How many are highly motivated?
  - Of those, how many tend to attrit despite the motivation?
  - How many are unmotivated? Did they start out unmotivated? Why are they unmotivated?
  - Of those, how many tend to attrit because of the lack of motivation?
  - Are any of the students unhappy about participating in this training?
  - Do you believe that these motivation issues are the cause for attrition?
- Recruiting
  - Are there special incentives for going to this AFSC?
  - Do students know what they're getting into?
  - Are most students here by choice?
  - Why do students choose this career field?
  - Is this good preparation for a civilian career later?
  - Do you believe that recruiting issues contribute to attrition?

- Washbacks
  - What blocks cause the most washback problems? Why?
  - Why are [trainees] successful the second time through, i.e., what changed?

Student Interview Questions

- Joining the Air Force
  - What were your reasons for joining the Air Force?
  - Where did you get your information about what the Air Force would be like, prior to actually joining?
  - Before you joined the Air Force, what did you expect it to be like?
  - Is being in the Air Force similar to what you thought it would be like?
  - How is it different?
  - How is it the same?
- Selection into your career training
  - What were you told during the recruiting process or during BMT about this AFSC?
  - If not nothing, does your experience match what you were told it would be?
  - What experiences or qualifications make someone suitable for this job?
  - Did you have a contract for this training when you joined the Air Force?
  - If not a contract, was this one of the jobs that you requested?
  - If not a job you selected, why do you think you were selected?
  - And, what do you think about the way the Air Force groups recruits into training programs?
- The training program
  - Is the training interesting?
  - Do you see this training as being useful in a career after the Air Force?
  - Is the training program too easy, difficult, [or] about right?
  - What is the most difficult part of the course?
  - Is there any way [the Air Force] could make the course easier?
  - Do you have enough time to study?
  - Do you have additional duties? How much time do they take?
  - What do you think are some of the reasons that trainees might do particularly well in the training course?
  - Have you heard of trainees having problems with the training course?
  - What do you think are some of the reasons that trainees might have difficulty with the training course?
  - What are your impressions of the phase program?
- Attrition
  - Have you heard of trainees voluntarily quitting the course or considering it?
  - What do you think are some of the reasons that trainees might voluntarily quit the training course?
  - What, if anything, do you think can be done to reduce the numbers of students dropping out of the course, both voluntarily or involuntarily?

APPENDIX B

Student Questionnaire

At the end of the interview, we administered the following written questionnaire (with the questions randomized) to recruits in the AGE, CCT, ATC, and PJ AFSCs. We modified the questionnaire slightly for interviews with operations intelligence, network intelligence, and linguist AFSCs. Items indicated by "DL" were deleted, and those indicated by "AL" were added for the intelligence and linguist AFSCs.

Questionnaire

- Do you know why you were selected for this job?
- What experiences or qualifications make someone suited for this job?
- Do you think you will make a career (spend 20+ years) in the Air Force?
- If you are thinking of leaving the Air Force without making it a career, at what point do you think you will leave?
- If you are thinking of leaving the Air Force currently, what are your potential reasons for leaving?

Using the following scale from 1 to 5, please circle the number that indicates how strongly you agree with each of the following statements:

1	2	3	4	5
Strongly disagree	**Disagree**	**Neither agree nor disagree**	**Agree**	**Strongly agree**

(DL) It is really important that I do well in this training.	1	2	3	4	5
I am not interested in this AFSC.	1	2	3	4	5
(DL) I'm just not cut out for this job.	1	2	3	4	5
You have to be really smart to do this job.	1	2	3	4	5
If I had known [that] the job would be like this, I never would have voluntarily agreed to go into this AFSC.	1	2	3	4	5
When trainees fail, it's because they don't like the job.	1	2	3	4	5
When trainees fail, it's because they are afraid of the dangers associated with the type of work that has to be done on the job.	1	2	3	4	5
(DL) I do not feel motivated to do well in this training.	1	2	3	4	5
(DL) You have to be really careful and responsible to do this job.	1	2	3	4	5

I'm proud to have this AFSC.	1	2	3	4	5
When trainees fail, it's because the training is physically too hard for them.	1	2	3	4	5
This job is completely different [from] what I expected before I started training.	1	2	3	4	5
I am well suited for this AFSC.	1	2	3	4	5
You have to pay attention to the details when you do this job.	1	2	3	4	5
(DL) I am doing better in training than most of the other trainees.	1	2	3	4	5
I know I will pass this training course.	1	2	3	4	5
(DL) I am doing really well in this training.	1	2	3	4	5
I would like to leave the Air Force.	1	2	3	4	5
When trainees fail, it's because their instructors have it in for them.	1	2	3	4	5
(DL) I am not doing very well in training.	1	2	3	4	5
I wish I had a different job.	1	2	3	4	5
This AFSC training is really hard.	1	2	3	4	5
I really wish I had a different job in the Air Force.	1	2	3	4	5
When trainees fail, it's because they aren't motivated.	1	2	3	4	5
My instructors are satisfied with my performance.	1	2	3	4	5
I am really motivated to do well in this training.	1	2	3	4	5
This AFSC was one of my top choices for jobs in the Air Force.	1	2	3	4	5
(DL) I am doing worse in training than most of the other trainees.	1	2	3	4	5
Most trainees think [that] the training is hard.	1	2	3	4	5
You have to be fearless to do this job.	1	2	3	4	5
It is really important to me that I pass this training.	1	2	3	4	5
When trainees fail, it's because the training is intellectually too hard for them.	1	2	3	4	5
I would rather have a different AFSC.	1	2	3	4	5
(AL) My squadron supports my job of learning another language.	1	2	3	4	5
(AL) My instructors support my job of learning another language.	1	2	3	4	5
(AL) Base facilities (library, study and testing rooms, etc.) sufficiently support my job of learning another language.	1	2	3	4	5
You have to be really physically strong and [in shape] to do this job.	1	2	3	4	5

APPENDIX C

Results of Questionnaire Data

Table C.1 shows the mean item scores for each subgroup of the survey sample (eliminated students, AFSC). Where a subgroup mean score is significantly higher or lower than that of the rest of the sample, the subgroup score is listed first, as greater than or less than the mean score for the remainder. The last two columns are Bonferroni analysis-of-variance (ANOVA) tests—the first comparing eliminated students with students still in the pipeline, the second comparing career fields.

It is important to emphasize that the results of these t-tests and Bonferroni ANOVA tests warrant cautious interpretation, for several reasons: First, the sample size is small, with a total of only 70 observations and fewer than ten observations within some of the groups. Second, survey participants were not randomly selected; rather, they were a convenience sample of trainees who happened to be available to fill out the questionnaire on the day of our visit; and third, the group of eliminated trainees was not representative of the initial five career fields being researched but were from only the PJ and EOD career fields. This biases the results in the "eliminated" group and biases the PJ and EOD groups in comparison to AGE, ATC, and CCT. Last, given the sheer number of hypothesis tests performed on the same data, we would expect to have some significant results at the 95-percent confidence level by chance alone.

Beginning with the column comparing the mean item scores between eliminated students and all others, the table shows that the means for eliminated students were significantly higher for two items relating to the perceived level of danger in the career field. Given that the eliminated students come from two of the objectively more dangerous career fields in our research, this is not terribly surprising. Had all the eliminated students come from AGE and ATC, the results would likely have been different. That the "eliminated" group would exhibit significantly greater agreement with a statement about wanting to leave the Air Force is not surprising, but the fact that, on average, these interviewees disagreed more strongly with the statement about the career field being completely different from what they had anticipated is somewhat unexpected. The most obvious explanation would be that these individuals knew ahead of time that they were being placed in a career field for which they lacked interest or aptitude, yet they did not express significantly stronger agreements with those items on the survey.

There is a pleasing consistency in the finding that students in the EOD course report both more pride in their career field and less desire to have a different career field, but it must be remembered that this group contained no eliminated students. The same phenomenon might account for stronger agreement with the statement about failure being linked to lack of motivation.

Table C.1
Results of the Analysis of Questionnaire Data: Explosive Ordnance Disposal, Pararescue, Combat Control, Aerospace Ground Equipment, and Air Traffic Control

Item	Eliminated (n = 9)	EOD (n = 20)	Other (n = 4)	PJ (n = 8)	CCT (n = 15)	AGE (n = 10)	ATC (n = 4)	ANOVA by AFSC	ANOVA by Elimination
It is really important that I do well in this training.	4.8	4.9	(4.5 < 4.9)*	4.9	4.8	4.7	5.0		
I am not interested in this AFSC.	1.2	1.4	(3 > 1.5)**	1.0	1.7	2.0	1.3		
I'm just not cut out for this job.	1.7	1.5	2.5	1.4	1.8	1.9	1.3		
You have to be really smart to do this job.	3.4	3.6	(2.5 < 3.5)*	3.1	(3.9 > 3.3)*	3.2	3.3		
If I had known [that] the job would be like this, I never would have voluntarily agreed to go into this AFSC.	1.2	1.7	2.5	(1 < 1.9)*	3.8	(2.3 > 1.7)*	1.3		
When trainees fail, it's because they don't like the job.	2.7	2.7	2.5	2.6	1.6	(3.6 > 2.7)**	3.0		
When trainees fail, it's because they are afraid of the dangers associated with the type of work that has to be done on the job.	(3.6 > 2.6)*	2.8	2.5	3.3	2.8	(2.1 < 2.8)*	3.3		F = 5.45*
I do not feel motivated to do well in this training.	1.4	1.7	2.3	1.2	2.8	2.2	1.0		
You have to be really careful and responsible to do this job.	4.8	4.5	4.8	4.8	1.8	4.7	4.5		
I'm proud to have this AFSC.	4.9	(4.9 > 4.5)*	4.3	5.0	4.5	(4.2 < 4.7)*	5.0		
When trainees fail, it's because the training is physically too hard for them.	2.8	(2.1 < 2.6)*	2.3	(3.4 > 2.8)**	(2.9 > 2.3)*	(1.8 < 2.5)*	3.3	F = 2.74*	
This job is completely different [from] what I expected before I started training.	(1.9 < 2.7)*	2.9	3.3	(1.9 < 2.7)*	2.3	2.9	2.3		F = 4.35*
I am well suited for this AFSC.	4.4	4.1	(2.5 < 4.1)**	(4.6 > 4.0)*	3.9	4.2	4.8	F = 2.53*	
You have to pay attention to the details when you do this job.	5.0	4.9	5.0	5.0	(4.6 < 4.9)*	4.8	5.0		

Table C.1—Continued

Item	Eliminated (n = 9)	EOD (n = 20)	Other (n = 4)	PJ (n = 8)	CCT (n = 15)	AGE (n = 10)	ATC (n = 4)	ANOVA by AFSC	ANOVA by Elimination
I am doing better in training than most of the other trainees.	3.4	3.4	3.8	3.5	3.4	3.5	4.0		
I know I will pass this training course.	3.6	3.8	4.0	4.1	(3.5 < 4.1)*	(4.6 > 3.8)**	4.5		
I am doing really well in this training.	3.9	3.7	3.5	4.1	3.4	4.1	4.3		
I would like to leave the Air Force.	(2.3 > 1.4)*	1.4	2.0	2.0	1.5	1.3	1.5		F = 5.61*
When trainees fail, it's because their instructors have it in for them.	2.2	2.0	2.3	1.9	1.9	(1.3 < 2.0)*	1.8		
I am not doing very well in training.	2.1	1.8	2.3	2.0	2.0	1.4	1.3		
I wish I had a different job.	1.6	(1.4 < 2.1)*	(4.0 > 1.7)**	1.3	2.0	2.3	1.0	F = 3.54**	
This AFSC training is really hard.	3.9	3.4	3.3	(4.4 > 3.4)**	(4.1 > 3.3)**	(2.6 < 3.7)**	2.8	F = 4.89**	
I really wish I had a different job in the Air Force.	1.9	1.8	(4.0 > 1.9)**	1.3	2.3	2.1	1.0	F = 2.95*	
When trainees fail, it's because they aren't motivated.	3.3	(3.5 > 2.8)*	(3.8 > 2.9)*	3.6	3.9	3.8	4.0		
My instructors are satisfied with my performance.	3.6	3.6	3.3	3.9	(3.3 < 3.7)*	3.9	3.8		
I am really motivated to do well in this training.	4.3	4.3	(3.3 < 4.3)*	4.5	4.1	4.4	4.5		
This AFSC was one of my top choices for jobs in the Air Force.	4.8	3.9	(2.8 < 4.2)*	(5 > 4.0)*	4.6	(3.5 < 4.3)*	4.8	F = 3.32*	
I am doing worse in training than most of the other trainees.	1.7	2.0	2.3	1.6	2.1	2.2	1.0		
Most trainees think [that] the training is hard.	4.3	3.9	4.3	4.3	4.3	(2.9 < 4.1)**	3.8	F = 4.43**	
You have to be fearless to do this job.	(3.9 > 3.2)*	3.3	3.0	3.6	3.6	(2.5 < 3.4)**	3.5		
It is really important to me that I pass this training.	4.2	4.6	(3.8 < 4.6)*	4.8	4.3	4.8	5.0		

Table C.1—Continued

Item	Eliminated (n = 9)	EOD (n = 20)	Other (n = 4)	PJ (n = 8)	CCT (n = 15)	AGE (n = 10)	ATC (n = 4)	ANOVA by AFSC	ANOVA by Elimination
When trainees fail, it's because the training is intellectually too hard for them.	2.8	(3.4 > 2.8)*	3.8	2.8	2.7	2.7	(2.0 < 3.0)*		
I would rather have a different AFSC.	1.9	1.7	(4.0 > 1.8)**	1.5	1.8	2.0	1.3	F = 2.77*	
You have to be really physically strong and [in shape] to do this job.	3.4	2.8	(1.5 < 3.3)**	(4.5 > 3.0)**	4.4	(2.0 < 3.4)**	3.5	F = 14.62**	

NOTE: * = $p < 0.05$. ** = $p < 0.01$. Where a subgroup mean score is significantly higher or lower than that of the rest of the sample, the subgroup score is listed first as greater than or less than the mean score for the remainder.

The "other" group, which consisted of an Airborne Warning and Control System trainee, a surgical-services trainee, an aviation resource management trainee, and an aerospace medical-services trainee, is too small and diverse from which to draw inferences.

Questionnaire respondents in the pararescue career field registered significantly more agreement with items avowing that their career field requires physical strength and is generally difficult. This result is consistent with the physically and psychologically rigorous training exercises for which the pararescue career field is renowned. Note that PJs were also significantly more positive about their suitability and preference for their career field. This result could be a reflection of the fact that almost all pararescue trainees are GTEP.

Results for the respondents from the CCT career field suggest that they believe more strongly than others surveyed that their career field requires strength, intelligence, and courage. When the statements were expressed in terms of the reasons that trainees fail in CCT, however, none of these traits appeared to be as critical. Significantly more pessimism about performing satisfactorily is not easily explained. Possibly, the instructors tend to use more negative reinforcement in this career field. With a mean of 4.6, the finding of lower agreement on the need for attention to detail is more a function of the nearly unanimous strong agreement with this item among other respondents than an indication that attention to detail is unimportant in the CCT career field.

Although the mean value (2.37) was still more a denial than an affirmation of the statement, "If I'd known [that] the job would be like this, I never would have gone into this AFSC," survey respondents from the AGE schoolhouse were significantly less vehement in their dissent. Furthermore, they registered significantly higher agreement with the statement linking failure to distaste for the work. Seven of ten respondents strongly believed that they would pass their current training course, while eight of ten disagreed or were neutral as to whether the training was difficult. The tendency of AGE survey respondents to view the career as unchallenging could be related to their indicating significantly less pride in their career field than others who took the survey.

Due to inexact labeling of the survey, the ATC respondents were mixed in with CCT respondents, further decreasing our ability to draw meaningful conclusions. This group indicated significantly more disagreement with the notion that motivation and insufficiency of intellect are reasons for failure.

When the survey responses between career fields are compared using ANOVA, there are significant differences in the perceived difficulty and desirability of the career field, especially on the importance of strength required for success. When eliminated students are compared with students still in their respective pipelines, differences emerge in the desire to leave the Air Force, the perceived relevance of fear as a factor in attrition, and the degree to which the career field met with expectations.

Table C.2 summarizes a similar analysis of the questionnaires administered to the trainees in the four career fields analyzed in the second year of study. Essentially, the linguist trainees indicated that they feel more challenged, less motivated, less proud, and less content with their career field than their counterparts in operations intelligence and network intelligence. Among the linguists surveyed, length of the training, stress, and better opportunities in civilian employment were cited as potential reasons for leaving the Air Force.

Table C.2
Results of the Analysis of Questionnaire Data: Operations Intelligence, Network Intelligence Analysis, and Middle East and Far East Linguist

Item	Operations Intelligence (n = 20)	Network Intelligence Analysis (n = 10)	Middle East/ Far East Linguist (n = 41)	ANOVA by AFSC
I am not interested in this AFSC.	1.7	(1.3 < 1.78)*	1.83	
You have to be really smart to do this job.	3.65	3.3	3.90	
If I had known [that] the job would be like this, I never would have voluntarily agreed to go into this AFSC.	(1.6 < 2.45)**	2.1	(2.54 > 1.77)**	F = 4.62**
When trainees fail, it's because they don't like the job.	(3 > 2.41)**	2.44	2.40	
When trainees fail, it's because they are afraid of the dangers associated with the type of work that has to be done on the job.	1.75	1.8	1.63	
I'm proud to have this AFSC.	(4.65 > 4.32)*	4.60	(4.25 < 4.63)*	
When trainees fail, it's because the training is physically too hard for them.	1.53	1.67	(1.95 > 1.57)*	
This job is completely different [from] what I expected before I started training.	(2.7 < 3.27)*	2.8	(3.39 > 2.73)**	
I am well suited for this AFSC.	4.25	4.1	4.00	
You have to pay attention to the details when you do this job.	4.75	(4.9 > 4.67)*	4.63	
I know I will pass this training course.	(4.55 > 3.88)**	(4.8 > 3.95)**	(3.66 < 4.63)**	F = 10.67**
I would like to leave the Air Force.	1.6	(1.3 < 1.72)*	1.78	
When trainees fail, it's because their instructors have it in for them.	1.9	(1.4 < 1.90)*	1.90	
I wish I had a different job.	1.8	1.89	(2.32 > 1.83)*	
This AFSC training is really hard.	(3.15 < 4.02)*	(2.3 < 4.02)**	(4.44 > 2.87)**	F = 26.67**
I really wish I had a different job in the Air Force.	1.95	1.9	2.35	
When trainees fail, it's because they aren't motivated.	3.55	3.44	3.24	
My instructors are satisfied with my performance.	4.0	4.38	3.93	
I am really motivated to do well in this training.	4.2	4.2	(3.83 < 4.2)*	
This AFSC was one of my top choices for jobs in the Air Force.	4.05	3.89	4.40	
Most trainees think [that] the training is hard.	(3.9 < 4.48)**	3.78	(4.63 > 3.86)**	F = 9.82**
You have to be fearless to do this job.	3.89	3.42	2.34	
It is really important to me that I pass this training.	4.58	4.56	4.34	

Table C.2—Continued

Item	Operations Intelligence (n = 20)	Network Intelligence Analysis (n = 10)	Middle East/ Far East Linguist (n = 41)	ANOVA by AFSC
When trainees fail, it's because the training is intellectually too hard for them.	3.4	3.11	3.24	
I would rather have a different AFSC.	1.95	2.0	2.24	
My squadron supports my job of learning another language.	2.39	2.38	2.80	
My instructors support my job of learning another language.	2.33	2.25	4.20	
Base facilities (library, study and testing rooms, etc.) sufficiently support my job of learning another language.	2.52	2.75	3.49	
You have to be really physically strong and [in shape] to do this job.	1.8	2.4	2.00	

NOTE: * = $p < 0.05$. ** = $p < 0.01$.

APPENDIX D

The Armed Services Vocational Aptitude Battery, Physical Ability and Stamina Test, and Federal Aviation Administration Air Traffic Selection and Training Test

The ASVAB is given to all recruits, whereas the PAST is only for PJ and CCT applicants.

Armed Services Vocational Aptitude Battery

The ASVAB has nine sections, the scores of which are combined in various ways by the various services. The individual sections are as follows:

- general science
- arithmetic reasoning
- word knowledge
- paragraph comprehension
- mathematics knowledge
- electronics information
- auto shop
- mechanical comprehension
- assembling objects.

The Air Force combines scores on these portions in various ways to create four measures collectively referred to as MAGE.

All nine of the career fields in this report except for AGE call for minimum scores in the general composite. The general score is composed of scores on the word knowledge, paragraph comprehension, and arithmetic reasoning portions of the ASVAB. In addition, EOD candidates are required to have minimum scores on the mechanical composite, which is made up of the three parts of the general score already mentioned plus mechanical comprehension. The AGE career field does not have a requirement for the general composite but uses scores on the mechanical composite and an electrical composite. This is a combination of scores on arithmetic reasoning, mathematics knowledge, electronics information, and general science. Individual MAGE measures do overlap because there are common components across the four individual measures.

Physical Ability and Stamina Test

In this section are the most-recent definitions of the PAST obtained from AFRS.

- Effective June 1, 2007, AFRS began administering a new PAST to qualify individuals for AFSCs 1C2X1 (CCT) and 1T2X1 (PJ). In order to ensure that applicants are fully qualified for these AFSCs, the recruiting service requires that they pass the following PAST prior to reservation and once again 30–45 days prior to entering active duty.
- All PJ and CCT applicants shipping from August 1, 2007, thru September 30, 2007, are administered a second PAST 30–45 days prior to their enter active duty (EAD) date. After a PJ or CCT applicant completes the second test, his squadron OPS must fax the results from this test to the Group OPS, and the Group OPS, in turn, needs to fax them to headquarters (HQ) AFRS. The results of these PASTs are used by the PJ and CCT indoctrination course to evaluate the fitness level of the new PJs and CCTs entering the pipeline. If the PJ or CCT applicant does not pass the second test, he is still allowed to ship on his current EAD date.
- Beginning on October 1, 2007, all PJ and CCT applicants are administered a second PAST 30–45 days prior to EAD. If a PJ or CCT applicant fails the second test, he is no longer eligible for the PJ or CCT career field and must be reclassified.
- The 1C2X1 and 1T2X1 career fields are open to male applicants only.

This test must be completed in a three-hour time frame, meeting the criteria listed below and in the order shown. Members are encouraged to complete the test in its entirety to determine weak and strong points in their physical condition. This test is comprised of seven events, and the member must pass every event. Failure of any event will result in the overall failure of the PAST.

The PAST requirements are designed to test for a minimum fitness level for entry into the CCT or PJ training pipeline. CCT and PJ candidates should continue to train throughout their cross-training application and the recruiting process, to exceed these minimums in order to enhance their chances of success.

The specific criteria are as follows:

- A11.1.1. 2 × 20-meter underwater swim. Provide the member three minutes of rest between the underwater tests. If the member surfaces or breaks the water surface during any portion of the swim, the test is stopped and considered a failure. Swimsuits and swim goggles or scuba mask are the only equipment items allowed. After the applicant completes the underwater swim, he or she is allowed a ten-minute rest before the next event.
- A11.1.2. 500-meter surface swim (maximum time limit is 14 minutes). This swim is conducted using the freestyle [front crawl], breaststroke, or sidestroke. The swim is continuous (nonstop). If the member stops any time during the swim, the test is stopped and considered a failure for the entire PAST. Swimsuit and goggles or scuba mask are the only equipment items allowed. After the member completes the surface swim, he or she is allowed a 30-minute rest prior to the next event.
- A11.1.3. 1.5-mile run (maximum time limit is ten minutes, 45 seconds). Physical-training clothes and good running shoes are the only required items. The run must be continuous (nonstop). If the member stops anytime during this run, the test is stopped and consid-

ered a failure. The member is given a ten-minute break prior to the next event. The test should be conducted on a measured running track.

- A11.1.4. Calisthenics.[1] Four calisthenic exercises are evaluated, each with specific time parameters and specific exercise-form mechanics. The member exercises to either muscle failure or time completion, whichever occurs first. The intent is to have the member do as many good-form repetitions in the time allotted or until muscle failure is reached. The member is allowed a three-minute rest between each calisthenic exercise.
- A11.1.5. Chin-ups and pull-ups (six-repetition minimum in one minute). Chin-ups and pull-ups are a two-count exercise. Starting position is hanging from a bar, palms facing toward or away from the candidate, with no bend in elbows. Hand spread is approximately shoulder-width apart. The member counts one then pulls the body up until the Adam's apple clears the top of the bar. On two, the member returns to starting position. Legs are allowed to bend but must not be kicked or manipulated to aid upward movement. If the candidate falls off, stops, or releases the bar, the exercise is terminated.
- A11.1.6. Sit-ups (45-repetition minimum in two minutes). Sit-ups are a two-count exercise. Starting position is with the member's back flat on the surface, fingers interlocked behind the head, head off the surface, and knees bent at approximately a 90-degree angle. Another individual may hold the individual's feet during the exercise. On the count of one, the member sits up so that the shoulders are directly above the hip or pelvis area or 90 degrees to the surface. On two, the member returns to the starting position. The exercise is continuous. If the member stops, the exercise is terminated. If the member's buttocks rise from the surface or his or her fingers are not interlocked behind his or her head during the repetition, the repetition is not counted.
- A11.1.7. Push-ups (45-repetition minimum in two minutes): Push-ups are a two-count exercise. Starting position is with hands shoulder-width apart, with arms straight and directly below the chest on the surface; the legs are extended, and the back and legs remain straight. On the count of one, the member lowers his or her chest until the elbows are bent at a 90-degree or lower angle. On two, he or she returns to the starting position. The only authorized rest position is the starting position. If the knees touch the ground, the exercise is terminated. The member may not raise his or her buttocks in the air, sag his or her middle to the surface, or raise any hand or foot from its starting position. If a hand or foot is raised, the exercise is terminated.
- A11.1.8. Flutter-kicks (45-repetition minimum in two minutes): Flutter-kicks are a four-count exercise. Starting position is the member lying flat on his or her back, with the feet and head approximately six inches off the surface. Hands are under the buttocks, with fists clenched to support the lower back. On the count of one, the member raises his or her left leg off the surface to approximately a 45-degree angle, keeping the right leg stationary. On two, the member raises the right leg off the surface to approximately a 45-degree angle, moving the left leg to the starting position. Counts three and four are repeats of the same movements. Legs must be straight, with toes pointing away from the body. If the member rests his or her legs on the surface or stops the exercise movement to rest, the exercise is terminated.

[1] Exercise form is strictly enforced during the PAST and in the training pipeline. Those repetitions done without proper form are not counted and are to the member's disadvantage.

Federal Aviation Administration Air Traffic Selection and Training

Eight components of the FAA AT-SAT are noted in this section. An extensive discussion of the development and validation of the test is in Ramos et al. (2001). The test components are as follows:

- Applied math. Solve problems involving distance, rate, and time relationships.
- Dials. Identify and read analog dials on an instrument panel.
- Angles. Specify degrees of angles in diagrams and vice versa.
- Analogy. Solve verbal and graphical analogies by applying rules derived from a pair of items to a second pair.
- Letter factory. Simulate a factory assembly line that manufactures letters A through D. This requires multiple, concurrent tasks involving learning rules and prioritization. This test is extremely complicated, and only a live demonstration would be sufficient to enable one to understand it.
- Scan. Monitor a field for objects that appear, travel for a short distance, and disappear.
- Air traffic scenarios. Maintain separation and control of aircraft in a low-fidelity simulation of a radar screen.
- Experience questionnaire. Respond to statements relating to past experiences on five-point scale indicating level of agreement.

Trainees' Perceptions of Injustice

In reviewing our notes taken during the interviews and focus groups, we identified some comments possibly indicative of perceived injustice or lack of fairness. Our concern is that individual students' perceptions of injustice could negatively affect satisfaction, performance, and attrition.

There is scholarly literature on perceived justice. Within this body of research, there is a taxonomy of types of justice and studies that connect negative consequences, such as those just mentioned, to perceptions of injustice. Among the instances of perceived injustice that we found, we identified violations of three kinds of perceived justice as defined in the literature (Colquitt, 2001; Greenberg, 1993).

- distributive justice: the perceived fairness of the allocation of resources
- procedural justice: the perceived fairness of the procedures used to determine the allocation of resources
- informational justice: the perceived fairness of the quality and amount of information used to explain allocation decisions and the procedures on which they are based.

Although these classifications are defined in terms of resource allocations, the concepts can be applied to trainees' experiences related to such areas as evaluation and standards, punishment for infractions of rules, job assignments, and information (Sheppard, Lewicki, and Minton, 1992, pp. 102–103).

Counts of Justice Issues in the Interviews and Focus Groups

In our notes from interviews and focus groups, we identified 65 comments that suggested possible perceived-justice issues. The comments are grouped into five categories. Table E.1 lists the categories and reports the frequency with which each was mentioned within each AFSC. The remainder of this appendix discusses the results displayed in Table E.1 and their implications. However, there are several limitations worth keeping in mind.

First, these analyses are based on comments indicative of justice issues as suggested by established findings in the literature (Greenberg and Colquitt, 2005). However, we did not explicitly measure perceived justice during this research, and we did not utilize established measures of perceived justice. Moreover, we did not have a representative sample of trainees.

Second, although the questions were structured, the open-ended nature of the responses made it possible for respondents to express or withhold justice concerns as they wished. There-

Table E.1
Number of Comments by Career Field Suggesting Possible Sources of Perceived Injustice

	Career Field						
Justice Issue	**PJ**	**EOD**	**AGE**	**ATC**	**CCT**	**Other**	**Total**
Recruiters provided information that was limited and/or inaccurate.	12	6	4	4	8	2[a]	36
Some recruits had less information than others about training/job assignments.	0	0	3	2	0	0	5
Different standards were used, or standards were applied unequally.	2	9	1	1	2	1[a]	16
Training evaluations were not indicative of what really was learned.	1	3	0	1	0	0	5
Because entire squadrons were punished, individuals who did nothing wrong suffered for the misdeeds of others.	0	0	1	2	0	0	3
Total	15	18	9	10	10	3[a]	65

SOURCE: First 75 student interviews conducted in 2008. An additional 79 student interviews conducted in 2009 were not included.

[a] One of each of the following AFSCs: 4N1XX, 1C5XX, and 1C0XX.

fore, failure to express concern about a particular source of injustice might not be indicative of respondents' beliefs that no such injustices occurred.

Third, none of the structured questions focused on justice per se, although some questions were more relevant to justice issues than to other issues. For example, we did ask questions regarding information imparted by recruiters. These questions yielded justice-relevant information, so it is not surprising that the highest number of justice-related comments are in this category. Thus, we cannot infer from the category totals a ranking on the basis of the amounts of perceived injustice associated with the categories.

Fourth, because the data were obtained in focus-group sessions, the participants heard one another's responses. As a result, the responses of one individual could prompt like responses by others. This appears to have occurred in some cases, thereby exaggerating the apparent popularity of some responses relative to what might have been observed had responses been made independently.

Fifth, potential injustices raised in the focus groups and interviews are perceptions expressed by participants, which might or might not reflect the actual fairness of the procedures in question.

Finally, the counts in Table E.1 are based on the observations of a single expert coder, which does admit the possibility of idiosyncratic bias.

Considering these points, caution should be exercised in assessing both the absolute and relative nature of the data in Table E.1. Nevertheless, several broad observations are worth noting. Issues of possible injustice in recruiting (the first two categories) and training processes are discussed separately, and recommendations for reducing perceived injustice follow.

Perceived Justice in Recruitment

Justice issues related to experiences in recruiting are counted in two categories. The first, which dominates the counts in the table, is about limited or inaccurate information provided by recruiters. The second, much less prevalent, is perceived injustices based on some recruits' belief that they had received less information than others had.

Recruiters Provided Information That Was Limited or Inaccurate

That this concern was the most frequently mentioned issue is not surprising since we asked specific questions about recruiting but not about any of the other justice issues listed in the table. Because information plays an important role in the making of career decisions (S. Schmidt, 2007), failure to obtain sufficient and accurate information could result in a perception of injustice. Receiving limited, incomplete, or inaccurate information about jobs, training, or assignments constitutes a violation of informational justice.

These perceptions of informational injustice appear to have two causes. Some trainees believed that recruiters oversold them on particular training options or made false promises because they were under pressure to fill certain positions. Others believed that recruiters had insufficient knowledge about the particular training options in which the recruits were most interested. These could be a result of recruiters' misplaced priorities due to a reward system that puts recruiters' interests above recruits' interests. Or, they could be due to inadequate training of recruiters that fails to ensure that recruiters are versed in a broad range of options and have good access to information to share with prospective recruits.

Some Recruits Had Less Information Than Others About Training and Job Assignments

Expressions of this issue were low in number overall and occurred only among AGE and ATC students. As recruits came to learn that some of their cohorts enjoyed the benefits of having higher-quality information on which to base their enlistment decisions, they were inclined to believe that they were underbenefited relative to these individuals with respect to this valuable resource. As such, they were victims of not only informational injustice but also distributive injustice.

This phenomenon is based on the existence of informal communication networks through which trainees compare their experiences with one another. When the comparisons are invidious in nature and an individual decides to escape the source of injustice by leaving the Air Force, others might be inclined to do the same. We noted three explicit mentions of this occurring.

As the initial faces of the Air Force, it is important that recruiters project messages of fairness by going out of their way to give recruits clear, thorough, and accurate depictions of training assignments.

Injustices Experienced During Training

Because they are relatively new to the Air Force, recruits undergoing training are inclined to scan their environments in search of information (Lind and van den Bos, 2002). In so doing, they make social comparisons that enable them to assess the appropriateness of their treatment relative to that of others in this new environment. Not surprisingly, this makes them hypervigilant to perceptions of injustice (Lind, 2001).

In particular, three distinct matters of injustice during the course of training were expressed. These relate to consistent application of standards, meaningful evaluations, and collective punishments.

Different Standards Were Used, or Standards Were Applied Unequally

This was the second–most widely cited justice issue overall, and EOD trainees accounted for more than half of the claims. Procedural justice requires that standards of treatment be applied in a consistent, even-handed manner (Leventhal, Karuza, and Fry, 1980).

The relatively high incidence of this category of injustice among EOD students could be related to the uniqueness of EOD training. As discussed previously, immediately after graduation from BMT, EOD trainees undergo a brief six-day screening course at Lackland. The actual, extensive training takes place at Eglin. Although Eglin is an Air Force base, the training for members of all services is conducted by the Navy.

Although the USAF trainees are relatively new to the military, representatives from the other services have more service experience, on average. And only the USAF students at Eglin are in a phase program. This could be the basis for claiming different standards, since students from other services are not subject to the standards of behavior imposed by the phase program. In some cases, the claims of differential treatment across the services might be explained by rank differences of trainees, and some trainees suggested that instructors favored students from the trainer's branch of service. This perceived favoritism might or might not be real.

Some trainees claimed that different instructors, even within the same base, used different standards. Such assessments are unlikely to be based on the principled observation that a procedural injustice occurred but rather on the more self-involved notion that they would have performed better had the luck of the draw assigned them to a more lenient instructor. Although the validity of this claim might be circumspect, the mere belief that such inequality between instructors exists is likely to prevail because it might give psychological comfort to those who are underperforming. Specifically, this belief serves as a psychological defense mechanism by allowing airmen to attribute their subpar performance externally, to "tough" instructors, instead of internally, to themselves (Cramer, 1991). As a result, poor performers are discouraged from acknowledging the need to take appropriate remedial action.

A belief expressed by some participants was that their superiors treated equally performing trainees differently and that they received undeservedly unfavorable treatment relative to others. This is inevitable within competitive, zero-sum situations, such as can be found in training classes, because trainees (1) are highly sensitive to negative evaluations and (2) have available to them only incomplete information about how evaluations are made.

Training Evaluations Were Not Indicative of What Really Was Learned

Concerns in this category were expressed in relatively low numbers, and mostly among EOD trainees.

Procedural justice requires that evaluations be based on valid information (Greenberg, 1986). Several trainees complained, however, that this was not the case, citing several interrelated factors. In particular, EOD trainees noted that they had inadequate time to study for exams because of the many obligations they had after class (e.g., to perform PT). This problem was exacerbated by two interrelated factors. First, because course material is classified, materials from the classroom could not be removed for study. Second, EOD training puts a high premium on memorization. As explained by instructors, the EOD job typically requires one to enter a classified area to study information about a particular explosive device, commit that information to memory (because the information is classified), and then apply the information from memory in the field. Thus, the imposition of conditions in which memorization is important to success might not be considered unfair insofar as it emulates conditions actually faced in the field. Because some trainees might not have recognized this, they could have perceived these procedures to be unfair. If this is true, this aspect of perceived injustice might easily be remedied by improving communication between trainers and trainees about the reasons for the restrictions.

Because Entire Squadrons Were Punished, Individuals Who Did Nothing Wrong Suffered for the Misdeeds of Others

This was the least widely cited justice issue and was noted only by one AGE and two ATC-CCT students. The actual prevalence of this concern is difficult to assess, however, because participants were not asked about it directly. Although there is no apparent reason to believe that this is a serious issue, it would be unwise to dismiss it entirely. To the extent that they could be part of a more general aspect of morale, issues of justice in the administration of punishment might be quite important.

The practice of punishing an entire squadron for an infraction committed by a particular airman would seem to be predicated on the belief that it promotes peer pressure to refrain from violating the rules (Wyatt and Gal, 1990). In civilian society, the practice of punishing a group for the infractions of an individual would be seen as a violation of distributive justice. In military organizations, this practice, and adverse reactions to it, reflects an ongoing set of opposing tensions between individualistic and collectivistic norms.

On the one hand, service members are differentiated with respect to their individual skills and accomplishments and are promoted on this basis. Under such conditions, punishing only individuals who break rules is most appropriate.

Additionally, airmen occasionally need to coordinate their efforts by working together as a team whose members make individual sacrifices for one another and for the greater good. In such situations, punishing entire groups is reasonable. Thus, the appropriateness of the practice is contingent upon the situation.

Recommendations for Promoting Justice

The sources of injustice noted in the previous section suggest several courses of action to curtail them in the future. Some might be more practical to implement than others. Although each of the following recommendations is aimed primarily at a particular source of injustice, together, the overall impact is likely to be felt broadly due to the beneficial cumulative effect from promoting a climate of fairness (Liao and Rupp, 2005). To the extent possible, these recommendations should be implemented at both the system level (e.g., by forming policies) and the individual level (e.g., by training personnel appropriately).

Explain Decisions Using Thorough Information

In several contexts, airmen indicated that they lacked information about how decisions were made. Such violations of informational justice were expressed at two points: (1) during recruitment (e.g., when the full gamut of career options was expressed in terms that were unclear or incomplete) and (2) during training (e.g., when the basis for course evaluations was unclear).

Sometimes, such informational blackouts might be intentional, such as when leaders prefer traditional hierarchical arrangements in which lower-level personnel are given only as much information as is necessary to perform the jobs immediately confronting them (Masten, 1995). A more enlightened, modern approach calls for sharing with subordinates the basis for decisions, especially decisions affecting them directly (Kouzes and Posner, 2006). This practice is beneficial from both an instrumental perspective (e.g., by teaching subordinates how decisions are made) and an interpersonal perspective (e.g., by treating airmen with dignity and respect; Lind and Tyler, 1988). The sharing of information in a traditionally hierarchical organization also is beneficial insofar as it is a status-neutralizing act (Greenberg, 1989) that helps promote the development of work teams.

Apply Consistent Standards in a Consistent Manner

Consistency also is a fundamental basis for perceived justice, and the practice of "treating equals as equals" is a key rule of fairness. As a tenet of distributive justice, consistency refers to the similar treatment of different people who are equivalent along key dimensions. This would be violated here if there were inconsistent standards across services, training classes, and individual instructors.

The quest to promote consistency requires coordinated efforts among policymakers across units. For example, if different instructors use different grading standards, a move toward uniformity would be wise. Not only would the resulting consistency remove a vexing source of procedural injustice, but it also would help ensure that appropriate standards are applied, thereby improving the quality of training.

It is important to note that it is not only the creation and implementation of consistent standards and practices that matters, but also knowledge of such efforts (Greenberg, 1990a). This requires publicity.

Base Course Grades on Performance

Among the most-cherished principles of learning and motivation is that trainees should be rewarded in proportion to their performance (Hume, 1995). However, to the extent that recruits believe that their course grades fail to reflect what they have learned, this principle would appear to be violated.

To some degree, this is a matter of managing perceptions. For example, it is necessary for EOD trainees to memorize large amounts of material without removing it from the classroom. Although trainees might consider this rule to be unfair because it does not tap their true capacity to understand the material, the rule is necessitated by the classified nature of the material. Moreover, because this procedure simulates field conditions in which memorization is necessary, its appropriateness is apparent to trainers. For trainees to share this judgment, however, it is necessary for them to be given thorough explanations that justify the procedures used. Such explanations promise to go a long way toward helping trainees appreciate the fairness of this assessment procedure.

If, however, evaluations (e.g., course grades) and performance (e.g., course accomplishments) are out of alignment, the evaluation procedure should be revised accordingly. One sure way of identifying valid bases for assessment is by tapping explicit course objectives. When material bearing on each objective is taught and assessed unambiguously, evaluations and performance are likely to be aligned (Eble, 1988).

Punish Only Individuals Who Break Rules Instead of the Entire Groups to Which They Belong

Group-wide punishment—just like group-wide reward—would be appropriate in work teams whose members are dependent on one another to achieve joint outcomes. This practice might be based on the assumption that it promotes social pressure within groups to meet standards. Although the extent to which this is accomplished is questionable, especially in the military (Wyatt and Gal, 1990), it is more likely that airmen will show signs of rejecting a system that punishes them unfairly. Rejection of such collective-punishment schemes is not unusual (Falomir-Pichastor et al., 2007).

Conclusion

Although promoting perceptions of fairness might be considered a desirable end in itself (Greenberg, 1990a), organizations whose members recognize the fairness of their leaders reap many tangible benefits, including a possible reduction in attrition.

Attrition can be caused by many factors (Hom, Roberson, and Ellis, 2008), making it difficult to isolate the impact of any one factor (for an analysis of attrition in the military, see U.S. General Accounting Office, 1998). However, given the considerable strength of the established negative relationship between justice perceptions and attrition reported in the literature (Finn and Lee, 1972; Schaubroeck, May, and Brown, 1994; Simons and Roberson, 2003), it is reasonable to suspect that perceptions of injustice could play a role in attrition among the airmen referenced here.

Additional desirable effects of increasing perceived justice also have been found. For example, high levels of perceived justice have been linked to high levels of certain types of job performance (Kamdar, McAllister, and Turban, 2006), high levels of morale and commitment (Ang, Van Dyne, and Begley, 2003; here, these would be seen in terms of reenlistment), and low levels of deviant, counterproductive behavior (e.g., theft; Greenberg, 1990b; Tomlinson and Greenberg, 2005). There are many good reasons for the Air Force to be aware of and responsive to matters of perceived injustice.

APPENDIX F

Analytical Approach

Predicting Success and Failure

Data sets for each of the nine AFSCs in the research were created by merging data from Defense Manpower Data Center (DMDC), AETC TTMS, HRRD, AAE data, and U.S. Military Entrance Processing Command (USMEPCOM) matched by social security number, which was scrambled by an algorithm to create a unique RAND identification number. An airman was included in the data set if he or she had any one of the subject AFSCs (at any skill level) as his or her current duty AFSC, control AFSC, primary, secondary, third, or fourth AFSC, or as a duty AFSC in any of the most recent six "history" duty assignments in the personnel record. AAE files from September 2001 to May 2009 were scanned to identify the relevant populations. For the majority of AFSCs, this encompassed trainees from 1998 to 2008.

EQ-i data were investigated for possible inclusion in the research, but some AFSCs did not have any EQ-i data, others had too few individuals with EQ-i data, and the ATC AFSC had EQ-i data only on apprentice-course graduates, which voided their value as predictive variables for researching attrition.

The nine data sets contained a total of 30,560 records, though there was some redundancy created by airmen who trained in more than one of the AFSCs being researched (e.g., several airmen who were unsuccessful in the CCT training were reclassified to ATC).

Within the data sets, airmen who successfully completed the apprentice course training were identified any of three ways: (1) by having a primary AFSC, duty AFSC, or duty history AFSC of the 3-level or higher (5-, 7-, or 9-level); (2) by having TTMS data for a 5-level course or above; or (3) by having a TTMS status code of GEE (graduated: equipment) for the final course in the 3-level training pipeline. Attrition from the apprentice course was coded for airmen who had a 1-level primary AFSC or a duty history indicating enrollment in the 3-level course but no duty history or primary AFSC indicating attainment of 3-level status or above and for airmen who had any TTMS status code indicating elimination from a course in the 3-level training pipeline (i.e., LAE = academic deficiency, LDE = disciplinary, LEE = self-elimination, LGE = separated, LIE = misconduct, LME = medical reason, LPE = prerequisite deficiency, LRE = performance deficiency, LTE = administrative reason, LUE = unsuitability, or LXE = other reason). We did not want to focus exclusively on academic attrition because, although elimination for these various reasons might have very different underlying causes, this is not necessarily the case.

Given the essentially binary outcome of success or failure in the training pipeline, attrition was modeled as a logistic regression of the form

$$\frac{P_i(pass)}{1-P_i(pass)} = \exp\left(\sum_{i=1}^{n} \beta_i X_i\right).$$

The covariates included in the model were age at the time of training, education, an interaction term for education and age, gender, race, ASVAB subtest scores, AFQT percentile, and whether the individual was a reservist or PS airman when he or she entered technical training. Indicator variables for each of the 50 states, the District of Columbia, and the territories were included to control for state-level effects. In consideration of the fact that CCT and PJ are elite, physically demanding career fields, for these AFSCs, we included height, weight, BMI, systolic and diastolic blood pressure, and an indicator for whether the airman was able to lift 100 pounds or more on the strength aptitude test. We omitted gender from the CCT and PJ models because women are not eligible for these career fields, although there is record of one woman who took the CCT 3-level training and passed.

Race was coded by consolidating pre- and post-2002 USAF race/ethnicity codes: "white" for "white/Caucasian," "black" for "black or African American," "native" for "American Indian" or "American Indian/Alaska Native," "asian" for "Asian," and "other" for "Other" before 2002, and for any of the myriad combinations of mixed ancestry from which applicants have been able to choose since 2002. Education was coded as a categorical variable with three levels: high school diploma or less, some college (but less than a four-year degree), and bachelor's degree or higher. Initially, an interaction term for education and age was included in the model, but it never substantially improved model goodness of fit.

For greater ease of interpretation, we use the relative-risk ratios for explanatory variables in our analysis of the binary pass/fail outcome. These were computed using a generalized linear model (specifically, a modified Poisson regression) with a logarithmic link function and robust standard errors. The coefficients of the variables are readily understood as estimates of the percentage difference in probability of passing with a one-unit change in the explanatory variable. The second data column in the table reports the test statistic.

Model goodness of fit was evaluated on the basis of McFadden's pseudo R^2 and on the result of the Hosmer-Lemeshow test, which compares the number of observed and expected cases of success and failure within each decile of the predicted probabilities. A significant chi-square statistic indicates model misspecification, such as nonlinearity in the right-hand variables or omitted-variable bias. The sensitivity and specificity of the logistic regression model were assessed by setting the threshold predicted probability for success to 0.5 and reporting the proportions of correctly classified observations.

An effort was made to predict final grade in the training course with the same set of covariates using a tobit regression with a lower limit of 70 and an upper limit of 100 for the lowest and highest possible passing scores, respectively. The tobit regression coefficients do not have a straightforward interpretation as effect sizes (i.e., the amount of increase in final grade per unit change in the covariate), so the emphasis is on the sign and statistical significance.

Air Traffic Control

Logistic Regression Correlations with Success

Data from 1,944 airmen who were enrolled in the ATC training pipeline between 2002 and 2007 were analyzed to identify correlates of attrition (other than attrition for medical reasons) from the AFSC (see Table F.1). For ATC, four variables were significant in predicting success among those selected for the training.

The group is already preselected for success in ATC. Additionally, ATC is primarily a volunteer-only job, although some students admitted to not understanding what the job entailed before volunteering. Therefore, we would not expect to see correlation among a group of variables already considered in selection.

Each one-point increase in arithmetic reasoning and mathematics knowledge scores increases success by 0.5 and 1.1 percent, respectively. An increase in the electronics information score causes a 0.5-percent decrease in success. The ability to lift 100 pounds or more should have no effect on ATC success (the entry requirement is to lift 40 pounds) yet improves success by 14 percent. Given the questionable aspect of the lifting variable, we recommend further

Table F.1
Logistic Regression Results for Success in Air Traffic Control (n = 1,944)

Variable	Risk Ratio	Z Statistic
Height	1.003	(0.11)
Weight	1.000	(0.06)
Systolic blood pressure	1.000	(0.43)
Diastolic blood pressure	1.001	(0.82)
BMI	1.000	(0.01)
Lift 100 lb. or more	1.139	(6.33)**
Age at enlistment in years	0.998	(0.45)
ASVAB general science	0.999	(0.45)
ASVAB arithmetic reasoning	1.005	(2.03)*
ASVAB word knowledge	1.003	(1.00)
ASVAB paragraph comprehension	1.003	(1.13)
ASVAB mathematics knowledge	1.011	(4.61)**
ASVAB electronics information	0.995	(2.45)*
ASVAB auto and shop information	0.998	(1.08)
ASVAB mechanical comprehension	0.997	(1.82)
ASVAB assembling objects	0.998	(1.04)

NOTE: * = significant at 5%. ** = significant at 1%. The McFadden's pseudo R^2 value of 0.04 as an index of goodness of fit indicates that the model is only a slight improvement on the null model. The p-value for the Hosmer-Lemeshow test for this model was 0.50, indicating no significant issue with goodness of fit.

analysis. The additional significance of four ASVAB subtest variables suggests a relook at the longtime practice of deciding vocational jobs on only four composite scores.

Tobit Regression Correlations with Final Grade

Because we did not have access to TTMS data for the ATC schoolhouse, we were unable to perform any tobit regression analysis of the effect that the model variables might have had on the final grade outcome.

Combat Control

Logistic Regression Correlations with Success

Data from 472 airmen who were enrolled in the combat control training pipeline between 2002 and 2007 were analyzed to identify correlates of attrition (other than attrition for medical reasons) from the AFSC (see Table F.2). For combat control, only two variables were significant in predicting success among those selected for the training.

The group is already preselected for success in combat control. Additionally, it is primarily a volunteer-only job, although some students admitted to not understanding what the job

Table F.2
Logistic Regression Results for Success in Combat Control (n = 472)

Variable	Risk Ratio	Z Statistic
Height	1.073	(0.57)
Weight	0.987	(0.51)
Systolic blood pressure	0.995	(1.18)
Diastolic blood pressure	0.006	(1.05)
BMI	1.105	(0.58)
Lift 100 lb. or more	1.197	(2.41)*
Age at enlistment in years	0.983	(0.98)
ASVAB general science	1.003	(0.44)
ASVAB arithmetic reasoning	1.000	(0.02)
ASVAB word knowledge	0.991	(1.02)
ASVAB paragraph comprehension	1.014	(1.74)
ASVAB mathematics knowledge	1.019	(2.20)*
ASVAB electronics information	0.995	(0.83)
ASVAB auto and shop information	1.001	(0.20)
ASVAB mechanical comprehension	0.998	(0.32)
ASVAB assembling objects	0.998	(0.33)

NOTE: * = significant at 5%. The McFadden's pseudo R^2 value of 0.04 indicates that the model is only a slight improvement over the null model; the Hosmer-Lemeshow test value of 0.004 indicates that there is also a model-specification issue.

entailed before they volunteered. Therefore, we would not expect to see correlation among a group of variables already considered in selection.

Each one-point increase in the mathematics knowledge score increased success by 1.9 percent. The mathematics knowledge subtest score probably relates to the ATC portion of the job. The ability to lift 100 pounds or more improves success by 20 percent. The entry requirement is to lift 70 pounds, and, since physical training dominates, it is not surprising to see greater success for those individuals who are stronger.

Tobit Regression Correlations with Final Grade

Final grades were also correlated with the same statistically significant variables in prediction success (see Table F.3). For final grade, lifting and mathematics knowledge were even more significant.

Table F.3
Tobit Regression Results for Final Score in Combat Control (n = 273)

Variable	dy/dx	Standard Error	Z Statistic	P > (z)	[95%	Confidence Interval]	Mean Value
Height	0.1494	0.6286	0.24	0.812	−1.083	1.381	69.655
Weight	−0.0351	0.1318	−0.27	0.790	−0.293	0.223	165.491
Systolic blood pressure	0.0039	0.0188	0.21	0.836	−0.033	0.041	125.788
Diastolic blood pressure	0.0192	0.0253	0.76	0.447	−0.030	0.069	71.850
BMI	0.3030	0.9105	0.33	0.739	−1.482	2.088	23.950
Lift 100 lb. or more	1.2856	0.4104	3.13**	0.002	0.481	2.090	0.227
Age at enlistment in years	0.0061	0.0710	0.09	0.932	−0.133	0.145	20.771
ASVAB general science	0.0191	0.0351	0.54	0.588	−0.050	0.088	55.876
ASVAB arithmetic reasoning	0.0020	0.0368	0.05	0.958	−0.070	0.074	56.550
ASVAB word knowledge	0.0105	0.0372	0.28	0.778	−0.062	0.083	54.711
ASVAB paragraph comprehension	0.0358	0.0329	1.09	0.275	−0.029	0.100	55.586
ASVAB mathematics knowledge	0.0974	0.0390	2.50**	0.013	0.021	0.174	57.231
ASVAB electronics information	−0.0630	0.0332	−1.90	0.058	−0.128	0.002	55.678
ASVAB auto and shop information	−0.0119	0.0277	−0.43	0.666	−0.066	0.042	54.429
ASVAB mechanical comprehension	0.0095	0.0296	0.32	0.749	−0.049	0.067	58.857
ASVAB assembling objects	−0.0173	0.0257	−0.67	0.501	−0.068	0.033	58.051

NOTE: dy/dx = discrete change of dummy variable from 0 to 1. P > (z) = the probability of the value occurring randomly. * = significant at 5%. ** = significant at 1%. y = E(TTMSfinalgrade2 | 70 < TTMSfinalgrade2 < 100) (predict, e(70,100)) = 83.31. McFadden's pseudo R^2 value of approximately 0.04 means that the model with predictors represents only a slight improvement over the model with no predictors.

Operations Intelligence

Logistic Regression Correlations with Success

Data from 1,350 airmen who were enrolled in the operations intelligence training pipeline between 2002 and 2007 were analyzed to identify correlates of attrition (other than attrition for medical reasons) from the AFSC (see Table F.4). For operations intelligence, only two variables were significant in predicting success among those selected for the training.

The group is already preselected for success in operations intelligence. Additionally, it is dominated by volunteers, although some students admitted to not understanding what the job entailed before they volunteered. Therefore, we would not expect to see correlation among a group of variables already considered in selection.

For this AFSC, each additional year of enlistment age reduces graduation success by 2.1 percent. It is not clear why this would make a difference, although the large amount of memorization might make it seem more tedious to older individuals.

Also, each point of the ASVAB arithmetic reasoning subtest decreases success by 0.7 percent. Again, this variable is part of the ASVAB general composite required for entry in the career field. A possible explanation is that, since this job is not technical but broad knowledge

Table F.4
Logistic Regression Results for Success in Operations Intelligence (n = 1,350)

Variable	Risk Ratio	Z Statistic
Height	0.981	(0.65)
Weight	1.005	(0.77)
Systolic blood pressure	0.998	(1.21)
Diastolic blood pressure	1.002	(0.79)
BMI	0.968	(0.78)
Lift 100 lb. or more	0.978	(0.47)
Age at enlistment in years	0.979	(2.85)**
ASVAB general science	0.997	(0.94)
ASVAB arithmetic reasoning	0.993	(2.05)*
ASVAB word knowledge	0.996	(1.10)
ASVAB paragraph comprehension	1.001	(0.32)
ASVAB mathematics knowledge	1.003	(0.80)
ASVAB electronics information	1.002	(0.71)
ASVAB auto and shop information	0.998	(0.88)
ASVAB mechanical comprehension	1.002	(0.57)
ASVAB assembling objects	1.005	(1.68)

NOTE: * = significant at 5%. ** = significant at 1%. The McFadden's pseudo R^2 value of 0.3 indicates that the model with these predictors is a slight improvement over the null model, and the Hosmer-Lemeshow p-value of 0.47 indicates no significant problem with model specification.

based, scores that are too high in technical areas might actually work against success because the individual is not motivated in his or her area of strength.

Tobit Regression Correlations with Final Grade

The factors that were significant in predicting success were not significant factors for predicting final grade in operations intelligence (see Table F.5). Age at enlistment would be significant at the 10-percent level. Mathematics knowledge, a subtest of the ASVAB, was not significant in predicting success, but, among those who passed the program (were successful), it was significant in predicting final grade.

Table F.5
Tobit Regression Results for Final Score in Operations Intelligence (n = 1,169)

Variable	dy/dx	Standard Error	Z Statistic	P > (z)	[95%	Confidence Interval]	Mean Value
Height	−0.3895	0.3058	−1.27	0.203	−0.989	0.210	67.723
Weight	0.0874	0.0668	1.31	0.190	−0.043	0.218	153.680
Systolic blood pressure	−0.0079	0.0169	−0.47	0.640	−0.041	0.025	122.639
Diastolic blood pressure	0.0159	0.0237	0.67	0.504	−0.031	0.062	74.339
BMI	−0.5742	0.4412	−1.30	0.193	−1.439	0.291	23.455
Lift 100 lb. or more	−0.2336	0.4198	−0.56	0.578	−1.056	0.589	0.151
Age at enlistment in years	−0.1012	0.0569	−1.78	0.075	−0.213	0.010	20.928
ASVAB general science	−0.0002	0.0323	−0.01	0.994	−0.064	0.063	56.953
ASVAB arithmetic reasoning	−0.0200	0.0353	−0.57	0.570	0.089	0.049	58.110
ASVAB word knowledge	0.0120	0.0372	0.32	0.746	−0.061	0.085	56.655
ASVAB paragraph comprehension	0.0529	0.0380	1.39	0.163	−0.022	0.127	57.581
ASVAB mathematics knowledge	0.0779	0.0332	2.34*	0.019	0.013	0.143	59.610
ASVAB electronics information	0.0303	0.0274	1.11	0.269	−0.023	0.084	54.102
ASVAB auto and shop information	−0.0221	0.0260	−0.85	0.395	−0.073	0.029	50.177
ASVAB mechanical comprehension	0.0160	0.0275	0.58	0.560	−0.038	0.070	56.244
ASVAB assembling objects	0.0497	0.0259	1.92	0.055	−0.001	0.100	57.774

NOTE: dy/dx = discrete change of dummy variable from 0 to 1. P > (z) = the probability of the value occurring randomly. * = significant at 5%. ** = significant at 1%. y = E(TTMSfinalgrade2 | 70 < TTMSfinalgrade2 < 100) (predict, e(70,100)) = 83.28. The McFadden's pseudo R^2 value of approximately 0.01 means that the model with predictors represents virtually no improvement over the model with no predictors.

Far East and Middle East Linguists

Logistic Regression Correlations with Success

Data from 1,002 airmen who were enrolled in the cryptological-linguist training pipeline between 2002 and 2007 were analyzed to identify correlates of attrition (other than attrition for medical reasons) from the AFSC (see Table F.6). For linguists, only three variables were significant in predicting success among those selected for the training.

The group is already preselected for success in linguistics. Additionally, it is all volunteers, although not necessarily for the particular language. For the most part, we would not expect to see correlation among a group of variables already considered in selection.

For this AFSC, the ability to lift 100 pounds (the requirement is 40 pounds) increases success by 6 percent. It is not clear why this would make a difference, though possibly the increased lift measures a more physically fit individual and, consequently, one more mentally fit as well.

Also, each point of the ASVAB mathematics knowledge subtest increased success by 0.7 percent. This variable is part of the ASVAB general composite required for entry in the

Table F.6
Logistic Regression Results for Middle East and Far East Linguist Trainees (n = 2,029)

Variable	Risk Ratio	Z Statistic
Height	1.000	(0.02)
Weight	0.999	(0.16)
Systolic blood pressure	1.001	(1.03)
Diastolic blood pressure	1.000	(0.07)
BMI	1.001	(0.05)
Lift 100 lb. or more	1.060	(2.68)**
Age at enlistment in years	1.005	(1.28)
ASVAB general science	1.000	(0.09)
ASVAB arithmetic reasoning	1.003	(1.14)
ASVAB word knowledge	1.000	(0.00)
ASVAB paragraph comprehension	1.003	(1.37)
ASVAB mathematics knowledge	1.007	(2.83)**
ASVAB electronics information	0.999	(0.34)
ASVAB auto and shop information	0.996	(2.53)*
ASVAB mechanical comprehension	0.999	(0.88)
ASVAB assembling objects	0.997	(1.91)

NOTE: * = significant at 5%. ** = significant at 1%. The McFadden's pseudo R^2 value of 0.03 indicates that the model with our predictors fits the data only slightly better than the null model. The Hosmer-Lemeshow p-value of 0.52 indicates that there is no serious problem with the model goodness of fit.

career field and suggests that more emphasis on the factor has a slight payoff. Likewise, a one-point increase in the ASVAB auto and shop information subtest decreased success by 0.3 percent, suggesting that experience in these activities does not indicate the right type of individual for linguistic training.

Tobit Regression Correlations with Final Grade

Four factors were significant in predicting success at Goodfellow language training (see Table F.7). The inclusion of these factors increases the explanation of variance by only 1 percent, making any conclusions dubious at best.

Table F.7
Tobit Regression Results for Middle East and Far East Linguists (n = 632)

Variable	dy/dx	Standard Error	Z Statistic	P > (z)	[95%	Confidence Interval]	Mean Value
Height	0.0356	0.3767	0.09	0.925	−0.703	0.774	67.816
Weight	−0.0330	0.0824	−0.40	0.689	−0.194	0.128	155.349
Systolic blood pressure	−0.0001	0.0218	0.00	0.996	−0.043	0.043	121.838
Diastolic blood pressure	0.0070	0.0316	0.22	0.824	−0.055	0.069	74.313
BMI	0.1530	0.5479	0.28	0.780	−0.921	1.227	23.645
Lift 100 lb. or more	2.5730	0.5336	4.82**	0.000	1.527	3.619	0.159
Age at enlistment in years	0.0675	0.0795	0.85	0.396	−0.088	0.223	21.112
ASVAB general science	0.0336	0.0421	0.80	0.425	−0.049	0.116	60.770
ASVAB arithmetic reasoning	0.0854	0.0491	1.74	0.082	−0.011	0.182	62.421
ASVAB word knowledge	−0.0288	0.0432	−0.67	0.506	−0.113	0.056	60.387
ASVAB paragraph comprehension	0.1379	0.0531	2.60**	0.009	0.034	0.242	60.618
ASVAB mathematics knowledge	0.1514	0.0497	3.05**	0.002	0.054	0.249	62.891
ASVAB electronics information	−0.0127	0.0337	−0.38	0.706	−0.079	0.053	58.021
ASVAB auto and shop information	−0.0663	0.0330	−2.01*	0.044	−0.131	−0.002	52.022
ASVAB mechanical comprehension	−0.0533	0.0333	−1.60	0.109	−0.118	0.012	60.310
ASVAB assembling objects	−0.0268	0.0343	−0.78	0.434	−0.094	0.040	61.538

NOTE: dy/dx = discrete change of dummy variable from 0 to 1. P > (z) = the probability of the value occurring randomly. * = significant at 5%. ** = significant at 1%. y = E(TTMSfinalgrade2 | 70 < TTMSfinalgrade2 < 100) (predict, e(70,100)) = 83.00. The McFadden's pseudo R^2 value of approximately 0.01 means that the model with our predictors is virtually no improvement over the model with no predictors.

Network Intelligence Analysis

Logistic Regression Correlations with Success

Data from 2,029 airmen who were enrolled in the network intelligence training pipeline between 2002 and 2007 were analyzed to identify correlates of attrition (other than attrition for medical reasons) from the AFSC (see Table F.8). For network intelligence, four variables were significant in predicting success among those selected for the training. These four values reduce the variation in results by 10 percent.

Age at enlistment has a positive effect on success in network intelligence (opposite of operations intelligence). A one-year increase in age improves the graduation rate by 0.8 percent. Likewise, a one-point increase in the ASVAB mathematics knowledge subtest improves the graduation rate by 0.8 percent. Additionally, one-point increases in assembling objects and paragraph comprehension have 0.5-percent increases in graduation rate. Network intelligence is more technical than operations intelligence, has less memorization, and has more interpretation and analysis. The additional ASVAB subtests suggest that the ASVAB general composite is not sufficient for selecting students for network intelligence training.

Table F.8
Logistic Regression Results for Network Intelligence Analysis (n = 1,002)

Variable	Risk Ratio	Z Statistic
Height	0.999	(0.06)
Weight	1.000	(0.21)
Systolic blood pressure	1.001	(1.04)
Diastolic blood pressure	0.999	(1.17)
BMI	1.003	(0.20)
Lift 100 lb. or more	1.029	(1.35)
Age at enlistment in years	1.008	(2.08)*
ASVAB general science	0.999	(0.29)
ASVAB arithmetic reasoning	1.000	(0.08)
ASVAB word knowledge	0.997	(1.09)
ASVAB paragraph comprehension	1.005	(2.18)*
ASVAB mathematics knowledge	1.008	(3.24)**
ASVAB electronics information	0.999	(0.77)
ASVAB auto and shop information	1.000	(0.08)
ASVAB mechanical comprehension	1.001	(0.53)
ASVAB assembling objects	1.005	(2.61)**

NOTE: * = significant at 5%. ** = significant at 1%. The McFadden's pseudo R^2 value of 0.10 indicates that the model with these predictors is a modest improvement over the null model, and the Hosmer-Lemeshow p-value of 0.06 is just shy of indicating a significant problem with misspecification.

Tobit Regression Correlations with Final Grade

All the same factors important for graduation plus two others (electronics information and arithmetic reasoning) were also significant in predicting final grade among graduates (see Table F.9).

Aerospace Ground Equipment

Logistic Regression Correlations with Success

Data from 2,310 airmen who were enrolled in the AGE training pipeline between 2002 and 2007 were analyzed to identify correlates of attrition (other than attrition for medical reasons) from the AFSC (see Table F.10). For AGE, three factors were significant in predicting success among those selected for the training. These factors (systolic blood pressure, diastolic blood

Table F.9
Tobit Regression Results for Final Score in Network Intelligence Analysis (n = 941)

Variable	dy/dx	Standard Error	Z Statistic	P > (z)	[95%	Confidence Interval]	Mean Value
Height	−0.0327	0.1590	−0.21	0.837	−0.344	0.279	67.932
Weight	0.0002	0.0346	0.01	0.996	−0.068	0.068	154.941
Systolic blood pressure	0.0242	0.0132	1.83	0.068	−0.002	0.050	122.516
Diastolic blood pressure	0.0034	0.0191	0.18	0.858	−0.034	0.041	73.870
BMI	−0.0200	0.1997	−0.10	0.920	−0.411	0.371	23.543
Lift 100 lb. or more	0.0449	0.3351	0.13	0.893	−0.612	0.702	0.167
Age at enlistment in years	0.1779	0.0589	3.02 **	0.003	0.063	0.293	20.685
ASVAB general science	0.0193	0.0254	0.76	0.447	−0.030	0.069	56.461
ASVAB arithmetic reasoning	0.0713	0.0316	2.26*	0.024	0.009	0.133	58.230
ASVAB word knowledge	−0.0238	0.0322	−0.74	0.461	−0.087	0.039	56.428
ASVAB paragraph comprehension	0.1160	0.0318	3.64**	0.000	0.054	0.178	57.343
ASVAB mathematics knowledge	0.1696	0.0275	6.17**	0.000	0.116	0.223	59.482
ASVAB electronics information	0.0635	0.0216	2.94**	0.003	0.021	0.106	54.254
ASVAB auto and shop information	−0.0356	0.0214	−1.67	0.096	−0.077	0.006	50.006
ASVAB mechanical comprehension	0.0022	0.0223	0.10	0.923	−0.042	0.046	55.866
ASVAB assembling objects	0.0458	0.0211	2.17*	0.030	0.004	0.087	57.753

NOTE: dy/dx = discrete change of dummy variable from 0 to 1. P > (z) = the probability of the value occurring randomly. * = significant at 5%. ** = significant at 1%. y = E(TTMSfinalgrade2 | 70 < TTMSfinalgrade2 < 100) (predict, e(70,100)) = 92.05. The McFadden's pseudo R^2 value of approximately 0.03 means that the model with predictors represents only a slight improvement over the model with no predictors.

Table F.10
Significant Logistic Regression Results for Graduation in Aerospace Ground Equipment (n = 2,310)

Variable	Risk Ratio	Z Statistic
Height	1.004	(0.18)
Weight	0.999	(0.24)
Systolic blood pressure	0.998	(2.00)*
Diastolic blood pressure	1.003	(2.08)*
BMI	1.016	(0.48)
Lift 100 lb. or more	0.972	(1.18)
Age at enlistment in years	0.970	(5.54)**
ASVAB general science	0.999	(0.66)
ASVAB arithmetic reasoning	1.000	(0.07)
ASVAB word knowledge	0.998	(0.91)
ASVAB paragraph comprehension	1.000	(0.18)
ASVAB mathematics knowledge	1.001	(0.27)
ASVAB electronics information	1.003	(1.71)
ASVAB auto and shop information	0.999	(0.48)
ASVAB mechanical comprehension	0.997	(1.65)
ASVAB assembling objects	1.000	(0.24)

NOTE: * = significant at 5%. ** = significant at 1%. The McFadden's pseudo R^2 value of 0.03 indicates that the model with these predictors is only a slight improvement over the null model. The p-value of the Hosmer-Lemeshow test for this model, which was 0.00, indicates a significant problem with the model goodness of fit.

pressure, and age at enlistment) are questionable due to a poor value for goodness of fit, which means that the model is not suited to the underlying distribution.

Tobit Regression Correlations with Final Grade

The model predicting final grade (see Table F.11) has statistically significant factors that are similar to the one for graduation but adds lifting 100 pounds and three ASVAB subtest scores (auto and shop information, electronics information, and assembling objects). Although the additional variables make sense, the model shows a 1-percent improvement in explaining the variation in final grades. Therefore, we do not trust the results to recommend any changes.

Pararescue

Logistic Regression Correlations with Success

Data from 801 airmen who were enrolled in the pararescue training pipeline between 2002 and 2007 were analyzed to identify correlates of attrition (other than attrition for medical rea-

Table F.11
Significant Tobit Regression Results for Final Score in Aerospace Ground Equipment (n = 1,602)

Variable	dy/dx	Standard Error	Z Statistic	P > (z)	[95%	Confidence Interval]	Mean Value
Height	0.198	0.321	0.62	0.538	−0.432	0.827	68.57
Weight	−0.036	0.070	−0.52	0.603	−0.173	0.101	157.49
Systolic blood pressure	−0.029	0.016	−1.78	0.075	−0.060	0.003	124.03
Diastolic blood pressure	0.062	0.022	2.79 **	0.005	0.018	0.105	74.09
BMI	0.334	0.473	0.71	0.480	−0.592	1.261	23.48
Lift 100 lb. or more	−1.797	0.367	−4.90**	0.000	−2.516	−1.078	0.16
Age at enlistment in years	−0.284	0.055	−5.20**	0.000	−0.392	−0.177	20.75
ASVAB general science	−0.008	0.030	−0.26	0.791	−0.067	0.051	54.02
ASVAB arithmetic reasoning	0.035	0.029	1.21	0.226	−0.022	0.092	53.82
ASVAB word knowledge	−0.043	0.035	−1.23	0.218	−0.112	0.026	52.92
ASVAB paragraph comprehension	0.012	0.031	0.38	0.707	−0.049	0.072	54.49
ASVAB mathematics knowledge	0.046	0.028	1.64	0.101	−0.009	0.102	55.03
ASVAB electronics information	0.064	0.024	2.70**	0.007	0.017	0.110	54.51
ASVAB auto and shop information	0.056	0.023	2.48*	0.013	0.012	0.101	54.19
ASVAB mechanical comprehension	−0.041	0.027	−1.52	0.128	−0.095	0.012	56.87
ASVAB assembling objects	0.058	0.023	2.50*	0.012	0.013	0.103	57.32

NOTE: dy/dx = discrete change of dummy variable from 0 to 1. P > (z) = the probability of the value occurring randomly. * = significant at 5%. ** = significant at 1%. y = E(TTMSfinalgrade2 | 70 < TTMSfinalgrade2 < 100) (predict e(70,100)) = 83.36. Final score in the AGE Apprentice course was not available for everyone in the AGE data. Therefore, the results in Table F.11 are based on a subset of the data in Table F.10. The McFadden's pseudo R^2 value of approximately 0.01 means that the model with predictors represents almost no improvement over the model with no predictors.

sons) from the AFSC (see Table F.12). For pararescue, four factors were significant in predicting success among those selected for the training and in explaining the 7-percent variation in results.

The group is already preselected for success in pararescue. Additionally, it is all volunteers. For the most part, we would not expect to see correlation among a group of variables already considered in selection.

The first three significant factors—systolic blood pressure, diastolic blood pressure, and lifting more than 100 pounds—are factors that are indicative of physical training. Those attributes are already measured in the PAST. Other studies have shown the importance of good physical conditioning prior to starting training. In this report, we specifically discuss the decline in physical fitness for these very fit individuals during BMT.

Table F.12
Significant Logistic Regression Results for Pararescue (n = 801)

Variable	Risk Ratio	Z Statistic
Height	1.408	(1.40)
Weight	0.936	(1.30)
Systolic blood pressure	1.028	(3.75)**
Diastolic blood pressure	0.966	(3.78)**
BMI	1.560	(1.23)
Lift 100 lb. or more	1.407	(2.46)*
Age at enlistment in years	0.996	(0.16)
ASVAB general science	1.008	(0.60)
ASVAB arithmetic reasoning	0.999	(0.05)
ASVAB word knowledge	1.005	(0.36)
ASVAB paragraph comprehension	0.998	(0.14)
ASVAB mathematics knowledge	1.059	(3.42)**
ASVAB electronics information	1.010	(0.93)
ASVAB auto and shop information	0.991	(0.82)
ASVAB mechanical comprehension	1.000	(0.03)
ASVAB assembling objects	0.998	(0.16)

NOTE: * = significant at 5%. ** = significant at 1%. Based on the McFadden's pseudo R^2 value of 0.07, the model is moderately higher than the likelihood of the null model. The Hosmer-Lemeshow p-value of 0.06 is just shy of indicating a serious issue with goodness of fit.

The fourth significant factor, the ASVAB subtest score of mathematics knowledge, is already included in the ASVAB composite general score. The requirement is quite low—a general score of 44—and this would suggest that there might be some advantage in a different composite for PJs instead of the current distribution of jobs among four ASVAB composites.

Tobit Regression Correlations with Final Grade

The model predicting final grade (see Table F.13) shows only one significant factor, mathematical knowledge. It does add an extra 8-percent explanation of the variation in final grades. This adds weight to the early result predicting graduation success.

Explosive Ordnance Disposal

Logistic Regression Correlations with Success

Data from 758 airmen who were enrolled in the EOD training pipeline between 2002 and 2007 were analyzed to identify correlates of attrition (other than attrition for medical reasons) from the AFSC (see Table F.14). For EOD, six factors were significant in predicting success among those selected for the training and in explaining the 6-percent variation in results.

Table F.13
Significant Tobit Regression Results for Final Score in Pararescue (n = 468)

Variable	dy/dx	Standard Error	Z Statistic	P > (z)	[95%	Confidence Interval]	Mean Value
Height	−0.3895	0.3058	−1.27	0.203	−0.989	0.210	67.723
Weight	0.0874	0.0668	1.31	0.190	−0.043	0.218	153.680
Systolic blood pressure	−0.0079	0.0169	−0.47	0.640	−0.041	0.025	122.639
Diastolic blood pressure	0.0159	0.0237	0.67	0.504	−0.031	0.062	74.339
BMI	−0.5742	0.4412	−1.30	0.193	−1.439	0.291	23.455
Lift 100 lb. or more	−0.2336	0.4198	−0.56	0.578	−1.056	0.589	0.151
Age at enlistment in years	−0.1012	0.0569	−1.78	0.075	−0.213	0.010	20.928
ASVAB general science	−0.0002	0.0323	−0.01	0.994	−0.064	0.063	56.953
ASVAB arithmetic reasoning	−0.0200	0.0353	−0.57	0.570	−0.089	0.049	58.110
ASVAB word knowledge	0.0120	0.0372	0.32	0.746	−0.061	0.085	56.655
ASVAB paragraph comprehension	0.0529	0.0380	1.39	0.163	−0.022	0.127	57.581
ASVAB mathematics knowledge	0.0779	0.0332	*2.34	0.019	0.013	0.143	59.610
ASVAB electronics information	0.0303	0.0274	1.11	0.269	−0.023	0.084	54.102
ASVAB auto and shop information	−0.0221	0.0260	−0.85	0.395	−0.073	0.029	50.177
ASVAB mechanical comprehension	0.0160	0.0275	0.58	0.560	−0.038	0.070	56.244
ASVAB assembling objects	0.0497	0.0259	1.92	0.055	−0.001	0.100	57.774

NOTE: dy/dx = discrete change of dummy variable from 0 to 1. P > (z) is the probability of the value occurring randomly. * = significant at 5%. ** = significant at 1%. y = E(TTMSfinalgrade2 | 70 < TTMSfinalgrade2 < 100) (predict e(70,100)) = 81.70. The McFadden's pseudo R^2 value of approximately 0.06 means that the model with predictors represents only a moderate improvement over the model with no predictors.

The ability to lift 100 pounds improved success by 37 percent (the requirement is 50 pounds). Five other ASVAB subtest scores affected graduation rates. One-point increases in arithmetic reasoning, paragraph comprehension, mathematics knowledge, and general science increased graduation rates by 1.8, 1.4, 1.4, and 1.1 percent, respectively. A one-point increase in electronics information *decreased* graduation rates by 0.9 percent. The latter result seems counterintuitive.

Tobit Regression Correlations with Final Grade

This population is a subset of the previous population, as it includes the final grades of only those who passed EOD. The model predicting final grade (see Table F.15) shows many of the same significant factors and explains 20 percent of the variation in final grades. Lifting 100 pounds and improvements in general science, arithmetic reasoning, and mathematics knowledge all improve final grade just as they improved graduation rate. Compared with

Table F.14
Significant Logistic Regression Results for Explosive Ordnance Disposal Graduation (n = 758)

Variable	Risk Ratio	Z Statistic
Height	0.935	(1.23)
Weight	1.013	(1.13)
Systolic blood pressure	1.002	(0.79)
Diastolic blood pressure	0.999	(0.40)
BMI	0.917	(1.08)
Lift 100 lb. or more	1.370	(7.70)**
Age at enlistment in years	1.005	(0.47)
ASVAB general science	1.011	(2.01)*
ASVAB arithmetic reasoning	1.018	(2.89)**
ASVAB word knowledge	0.991	(1.41)
ASVAB paragraph comprehension	1.014	(2.34)*
ASVAB mathematics knowledge	1.014	(2.65)**
ASVAB electronics information	0.991	(2.24)*
ASVAB auto and shop information	1.003	(0.87)
ASVAB mechanical comprehension	1.005	(1.01)
ASVAB assembling objects	1.000	(0.08)

NOTE: * = significant at 5%. ** = significant at 1%. Because of the McFadden's pseudo R^2 value of 0.06, the model appears to fit the data moderately better than the null model does. The Hosmer-Lemeshow p-value of 0.06 is just shy of indicating a serious issue with goodness of fit.

paragraph comprehension, which increased graduation rate, word knowledge better predicts final grade.

Age at enlistment is a new variable that did not predict success but does help predict final grade. This is consistent with a body of work discussed in the main report regarding the importance of maturity.

Finally, electronics information is a negative predictor of final grade as it was in predicting graduation. It is not clear why that might be true.

Table F.15
Significant Tobit Regression Results for Final Score in Explosive Ordnance Disposal (n = 524)

Variable	dy/dx	Standard Error	Z Statistic	P > (z)	[95%	Confidence Interval]	Mean Value
Height	−0.1176	0.1055	−1.11	0.265	−0.324	0.089	69.054
Weight	0.0193	0.0225	0.86	0.391	−0.025	0.063	160.441
Systolic blood pressure	0.0005	0.0050	0.11	0.914	−0.009	0.010	124.048
Diastolic blood pressure	0.0033	0.0073	0.46	0.645	−0.011	0.018	74.202
BMI	−0.1380	0.1544	−0.89	0.371	−0.441	0.165	23.600
Lift 100 lb. or more	1.2566	0.2248	5.59**	0.000	0.816	1.697	0.298
Age at enlistment in years	0.0528	0.0262	2.02*	0.044	0.001	0.104	20.455
ASVAB general science	0.0247	0.0114	2.17*	0.030	0.002	0.047	58.504
ASVAB arithmetic reasoning	0.0331	0.0135	2.44*	0.015	0.007	0.060	59.512
ASVAB word knowledge	−0.0128	0.0121	−1.06	0.290	−0.037	0.011	57.147
ASVAB paragraph comprehension	0.0091	0.0112	0.82	0.414	−0.013	0.031	57.884
ASVAB mathematics knowledge	0.0247	0.0111	2.24*	0.025	0.003	0.046	59.634
ASVAB electronics information	−0.0271	0.0095	−2.86**	0.004	−0.046	−0.009	57.418
ASVAB auto and shop information	−0.0019	0.0077	−0.24	0.807	−0.017	0.013	55.181
ASVAB mechanical comprehension	0.0021	0.0092	0.22	0.823	−0.016	0.020	60.155
ASVAB assembling objects	−0.0078	0.0087	−0.90	0.368	−0.025	0.009	59.063

NOTE: dy/dx = discrete change of dummy variable from 0 to 1. P > (z) = the probability of the value occurring randomly. * = significant at 5%. ** = significant at 1%. y = E(TTMSfinalgrade2 | 70 < TTMSfinalgrade2 < 100) (predict, e(70,100)) = 84.74. The McFadden's pseudo R^2 value of approximately 0.20 indicates that the model with predictors represents some improvement over the model with no predictors.

APPENDIX G

Aerospace Ground Equipment Personnel Mathematical Training Model

Given the high washback rate and low attrition rate of AGE, we developed a model to test various washback and attrition scenarios and compare the relative cost change between the scenarios.

We examine the two costs using a mathematical model. The model makes the following assumptions:

$$TC_i = sum(O \,\&\, M_i + S_i + I_i),$$

where TC_i is the total cost of block *i*; *O&M* is the operational and maintenance cost of block *i*; S_i is the student cost of block *i*; and I_i is the instructor cost of block *i*.

$$I_i = f(\text{number of instructors, length of the block, composite military pay per duty day}),$$

where composite pay is determined from Air Force Instruction (AFI) 65-503, Tables 19-2 and 47.

$$\begin{aligned}
S_i &= f(\text{length of block, composite pay per duty day}),\\
O \,\&\, M_i &= f(\text{length of block, O\&M cost per day}),\\
\text{AGE training program cost per student} &= \$39{,}690 \text{ (AFI 65-503, Tables 18-1A and 47)},\\
\text{BMT cost per student} &= \$9{,}333 \text{ (AFI 65-503, Tables 7-1 and 47)},\\
\text{AGE technical-training course cost} &= \text{AGE training program cost per student}\\
&\quad -\text{BMT cost per student}\\
&= \$30{,}357,\\
\text{All pay-related costs} &= \left(\frac{sum(I_i)}{\text{average class size}}\right) + sum(S_i) \text{ for all blocks}\\
&= \$21{,}986.\\
\text{Therefore, } O \,\&\, M_i \text{ per student} &= \text{AGE technical-training course cost}\\
&\quad -\text{all pay-related costs}\\
&= \$30{,}357 - \$21{,}986 = \$8{,}371.\\
\text{Therefore, O\&M per day} &= \$85.
\end{aligned}$$

Using these calculations, historical attrition, and washback rates by block, we are able to construct Figure G.1. The vertical axis is the increased costs over a 0-percent washback, 0-percent attrition case. The 361 TRS has reduced its attrition rate to 4.4 percent while seeing washbacks increase to 30 percent. Costwise, we learned that one person attriting costs the same as 2.2 washbacks. A person attriting has a 0-percent chance of graduating, whereas washbacks graduate at an 80.6-percent rate. The 2.2:1 ratio implies that, if AGE cut its washback rate in half (down to 15 percent) by eliminating more academically deficient personnel, the attrition rate could increase to 10.9 percent without any increase in costs. Figure G.1 illustrates this dynamic in that the cost increase over baseline with a 5-percent attrition rate and a 30-percent washback rate (about $460,000) is greater than the cost increase with a 10-percent attrition rate and a 15-percent washback rate (about $400,000).

We conclude that the AGE philosophy of allowing high washbacks is a cost-effective strategy for reducing attrition and training costs.

Figure G.1
Increased Training Costs by Washback Rate and Attrition Rate

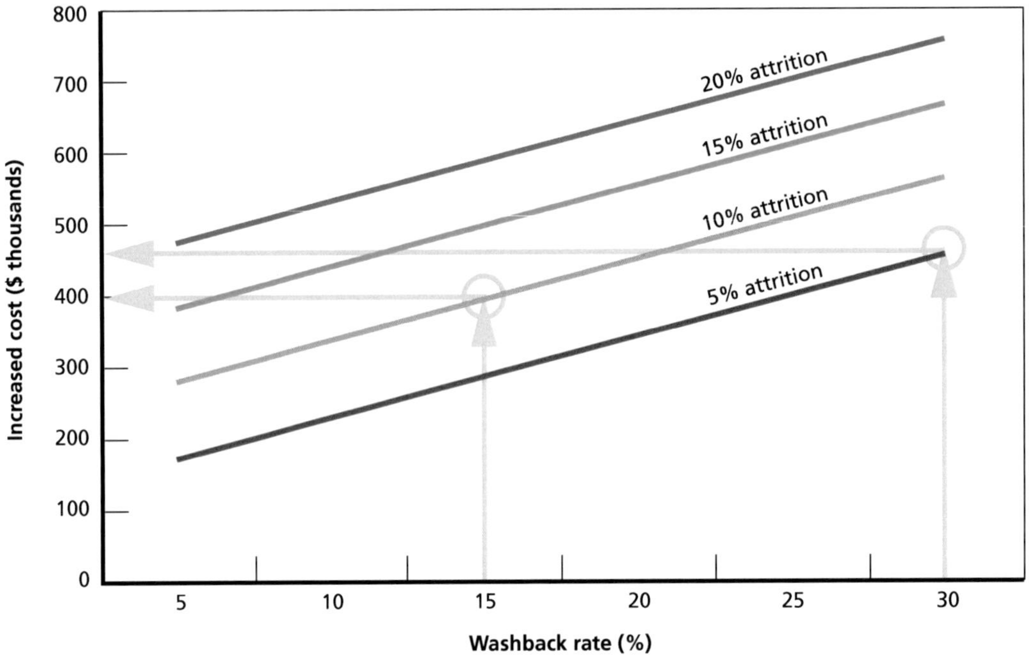

RAND *TR955-G.1*

APPENDIX H
Length of the Technical Training Program

To see whether there is a relationship between lengths of training programs and attrition rates, we analyzed data from approximately 200 distinct course pipelines over ten years, providing 1,407 data points.[1] Figure H.1 is a plot of percentage attrition versus length of the training pipeline. A linear regression produced a weighting of an 0.11-point increase in the attrition percentagc per week. Although "significant" in the specialized sense used by statisticians, the linear model accounts for only 3 percent of the variability in the observed attrition rates. When the 0.11 percent per week is extrapolated to 52 weeks, the result is 10 percent per year, which is approximately the attrition rate of first-term airmen.

The relationship shown in Figure H.1 explains very little of the variation. We tested the relationship for outlier effect by eliminating any course with more than 30-percent attrition, and there was little change in the factors. We also added the variable of ASVAB requirement as a measure of training difficulty. In this case, "training difficulty" was not significant in predicting attrition. Although not conclusive, it is true that longer programs, which generally have higher attrition, have greater average attrition per week of the program.

---

1 Originally, there were 1,790 data points. We reduced the data points to 1,407 after eliminating AFSCs with five or fewer students.

Figure H.1
Relationship Between Length of the Training Pipeline and Attrition Rate

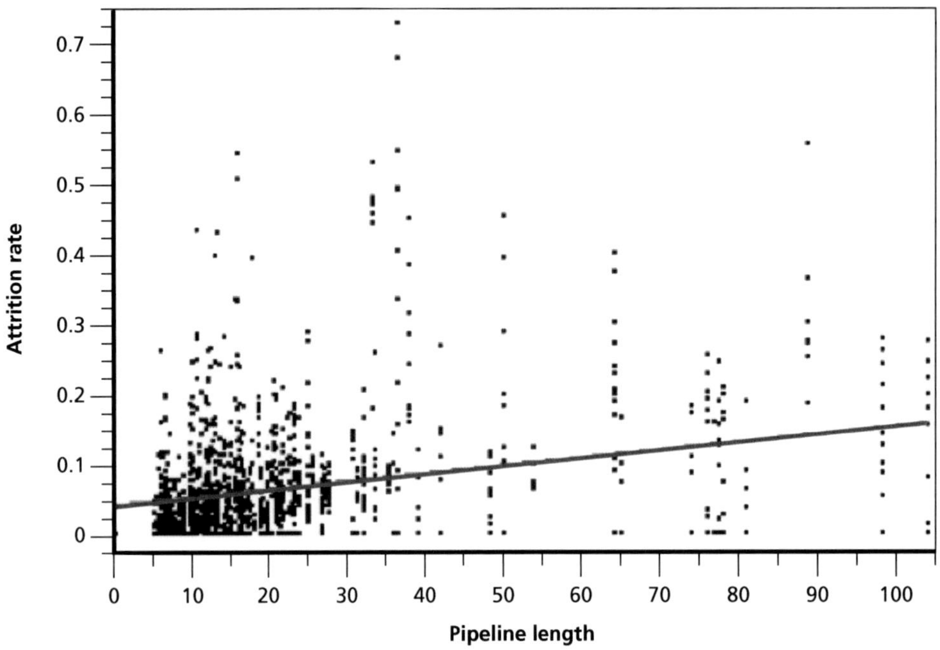

NOTE: Attrition rate = 4.27% + 0.11% per week.

RAND *TR955-H.1*

Bibliography

2AF—*See* 2nd Air Force.

2nd Air Force, "August 2010 RAND Report: Reducing Attrition in Selected Air Force Training Pipelines," memorandum, October 7, 2010.

37th Training Group, *Course Chart Pararescue (PJ)/Combat Rescue Officer (CRO) Indoctrination*, Lackland Air Force Base, Texas, L3AQR1T231 0P1A L3OQR13D1A 0P1A, August 2005a.

———, *Course Chart Combat Control Orientation Course*, Lackland Air Force Base, Texas, L3AQR1C231 0C0A L3OQR13D1B 0C0A, October 17, 2005b.

———, *Plan of Instruction (Technical Training) Combat Control Orientation Course*, Lackland Air Force Base, Texas, L3AQR1C231 0C0A L3OQR13D1B 0C0A, October 17, 2005c.

———, *Training Plan Explosive Ordnance Disposal Preliminary*, Keesler Air Force Base, Miss., L3AQRE3831 0E0B, January 22, 2007.

———, *Plan of Instruction (Technical Training) Pararescue (PJ)/Combat Rescue Officer (CRO) Indoctrination Course*, Lackland Air Force Base, Texas, L3AQR1T231 0P1A L3OQR13D1A 0P1A, January 10, 2008.

81st Training Group, *Plan of Instruction Air Traffic Control Fundamentals*, Keesler Air Force Base, Miss., E3AQR1C131 00FB, June 6, 2005a.

———, *Plan of Instruction Air Traffic Control Radar Apprentice*, Keesler Air Force Base, Miss., E3ABR1C131 00RB, June 7, 2005b.

———, *Plan of Instruction Air Traffic Control Tower Apprentice*, Keesler Air Force Base, Miss., E3ABR1C131 00TB, June 7, 2005c.

———, *Course Chart Air Traffic Control Radar Apprentice*, Keesler Air Force Base, Miss., E3ABR1C131 00RB, June 7, 2005d.

———, *Course Chart Air Traffic Control Tower Apprentice*, Keesler Air Force Base, Miss., E3ABR1C131 00TB, August 7, 2005e.

———, *Course Chart Combat Control Operator Fundamentals*, Keesler Air Force Base, Miss., E3AQR1C131 00FB, August 16, 2007.

———, *Course Chart Combat Control Operator Fundamentals*, Keesler Air Force Base, Miss., E3AQR1C231 00FC, January 30, 2008a.

———, *Course Chart Combat Control Operator Tower*, Keesler Air Force Base, Miss., E3AQR1C231 00TB, January 30, 2008b.

———, *Plan of Instruction (Technical Training) Combat Control Fundamentals*, Keesler Air Force Base, Miss., E3AQR1C231 00FC, January 30, 2008c.

———, *Plan of Instruction (Technical Training) Combat Control Operator*, Keesler Air Force Base, Miss., E3AQR1C231 00TB, January 30, 2008d.

82nd Training Group, *Plan of Instruction (Technical Training) Aerospace Ground Equipment Apprentice*, Sheppard Air Force Base, Texas, J3ABR2A632 046A, July 19, 2006a.

———, *Course Chart Aerospace Ground Equipment*, Sheppard Air Force Base, Texas, J3ABR2A632 046A, October 23, 2006b.

342nd Training Squadron, "Combat Control FY05/FY06/FY07 Production Analysis," briefing, 342 Training Squadron, Lackland Air Force Base, Texas, February 11, 2008.

361 Training Squadron/Deputy of Operations, "Production Analysis and Reparations in RG," memorandum for commander, 361 Training Squadron, Sheppard Air Force Base, Texas, January 3, 2008.

Abram, Stephen, and Judy Luther, "Born with the Chip," *Library Journal*, May 1, 2004. As of April 26, 2011: http://www.libraryjournal.com/article/CA411572.html

Ackerman, Phillip L., "Predicting Individual Differences in Complex Skill Acquisition: Dynamics of Ability Determinants," *Journal of Applied Psychology*, Vol. 77, No. 5, October 1992, pp. 598–614.

Ackerman, Phillip L., and Ruth Kanfer, "Integrating Laboratory and Field Study for Improving Selection: Development of a Battery for Predicting Air Traffic Controller Success," *Journal of Applied Psychology*, Vol. 78, No. 3, June 1993, pp. 413–432.

AETC—*See* Air Education and Training Command.

AETC/A3—*See* Air Education and Training Command Directorate of Intelligence, Operations, and Nuclear Integration.

AETC Instruction 36-2216—*See* Headquarters Air Education and Training Command (2004).

AF/A1—*See* Headquarters Air Force Personnel Directorate.

AFAA—*See* Air Force Audit Agency.

AFMAN 36-2108—*See* Air Force Personnel Center (2004).

AFPC—*See* Air Force Personnel Center.

Air Education and Training Command, *On Learning: The Future of Air Force Education and Training*, Randolph Air Force Base, Texas, January 30, 2008. As of April 26, 2011: http://www.aetc.af.mil/library/whitepaper.asp

Air Education and Training Command Directorate of Intelligence, Operations, and Nuclear Integration, "Technical Training Pipelines," c. 2007.

Air Force Audit Agency, *Hard to Fill Career Fields*, F2008-0008-FD4000, August 13, 2008.

Air Force Personnel Center, *Enlisted Classification*, Air Force Manual 36-2108, Randolph Air Force Base, Texas, October 31, 2004.

———, *Air Force Enlisted Classification Directory (AFECD)*, Randolph Air Force Base, Texas, December 2010.

Ang, Soon, Linn Van Dyne, and Thomas M. Begley, "The Employment Relationships of Foreign Workers Versus Local Employees: A Field Study of Organizational Justice, Job Satisfaction, Performance, and OCB," *Journal of Organizational Behavior*, Vol. 24, No. 5, August 2003, pp. 561–583.

Aronson, Julie, Joy Zimmerman, and Lisa Carlos, *Improving Student Achievement by Extending School: Is It Just a Matter of Time?* San Francisco, Calif.: WestEd, 1999.

Bates, Mark J., *A Risk Factor Model Predicting the Relationship Between Stress and Performance in Explosive Ordnance Disposal (EOD) Training*, Wright-Patterson Air Force Base, Ohio: Air Force Institute of Technology, October 7, 2002. As of April 26, 2011: http://handle.dtic.mil/100.2/ADA410705

Baumgartner, Neal, "Fit Flight and Physical Performance Data," email summarizing results of unpublished work, 342nd Training Squadron, August 19, 2008.

Belenky, Gregory, Shabtai Noy, and Zahava Solomon, "Contributions in Military Studies," in Gregory Belenky, ed., *Contemporary Studies in Combat Psychiatry*, New York: Greenwood Press, 1987, pp. 11–20.

Blumberg, P., and Flaherty, J. A., "The Influence of Noncognitive Variables in Student Performance," *Journal of Medical Education*, Vol. 60, No. 9, September 1985, pp. 721–723.

Bongiovi, Lt. Col. Thomas, "EOD Pipeline Repair," briefing, 366th Training Squadron, Detachment 3, Eglin Air Force Base, Fla., November 16, 2007.

———, "366 Training Squadron Detachment 3," briefing, 366th Training Squadron, Detachment 3, Eglin Air Force Base, Fla., January 29, 2008a.

———, "EOD School Overview," briefing, 366th Training Squadron, Detachment 3, Eglin Air Force Base, Fla., January 29, 2008b.

Breeding, Marshall, "Technology for the Next Generation," *Computers in Libraries*, Vol. 26, No. 10, November–December 2006, pp. 28–30. As of April 26, 2011:
http://www.librarytechnology.org/ltg-displaytext.pl?RC=12342

Buddin, Richard, *Success of First-Term Soldiers: The Effects of Recruiting Practices and Recruit Characteristics*, Santa Monica, Calif.: RAND Corporation, MG-262-A, 2005. As of April 26, 2011:
http://www.rand.org/pubs/monographs/MG262.html

Bundy, Edwin, and Roderick Sims, "Commonalities in an Uncommon Profession: Bomb Disposal," *Proceedings Ascilite Singapore 2007*, 2007. As of April 26, 2011:
http://www.ascilite.org.au/conferences/singapore07/procs/bundy.pdf

Burt, Diana Byrd, Mary Jo Zembar, and George Niederehe, "Depression and Memory Impairment: A Meta-Analysis of the Association, Its Pattern, and Specificity," *Psychological Bulletin*, Vol. 117, No. 2, March 1995, pp. 285–305.

Campbell, Donald, Kathleen Campbell, and James W. Ness, "Resilience Through Leadership," in Brian J. Lukey and Victoria Tepe, eds., *Biobehavioral Resilience to Stress*, Boca Raton, Fla.: CRC Press, 2008, pp. 57–90.

Carpenter, Kristin, "Tip: Providing Answers Easily and Effectively on the Web," *Enrollment Marketing Best Practices*, Vol. 1, No. 16, November 14, 2006.

Carretta, Thomas R., "USAF Enlisted Air Traffic Controller Selection," briefing, Warfighter Interfaces Division, Wright-Patterson Air Force Base, Ohio, June 5, 2007.

Carretta, Thomas R., and Raymond E. King, "Improved Military Air Traffic Controller Selection Methods as Measured by Subsequent Training Performance," *Aviation, Space, and Environmental Medicine*, Vol. 79, No. 1, January 2008, pp. 36–43.

Carretta, Thomas R., and Frederick M. Siem, "Determinants of Enlisted Air Traffic Controller Success," *Aviation, Space, and Environmental Medicine*, Vol. 70, No. 9, September 1999, pp. 910–918.

Center for Strategic and International Studies, *American Military Culture in the 21st Century*, Washington, D.C., 2001.

Cobb, Bart B., "Problems in Air Traffic Management II: Prediction of Success in Air Traffic Controller School," *Aerospace Medicine*, June 1962, pp. 702–713.

———, "Problems in Air Traffic Management V: Identification and Potential of Aptitude Test Measures for Selection of Tower Air Traffic Controller Trainees," *Aerospace Medicine*, Vol. 35, No. 11, November 1964, pp. 1019–1027.

———, *A Comparative Study of Air Traffic Trainee Aptitude-Test Measures Involving Navy, Marine Corps and FAA Controllers*, Washington, D.C.: U.S. Department of Transportation, Federal Aviation Administration, Office of Aviation Medicine, AM68-14, September 1968. As of April 26, 2011:
http://purl.access.gpo.gov/GPO/LPS80031

Cobb, Bart B., and John J. Matthews, "Proposed New Test for Aptitude Screening of Air Traffic Controller Applicants," *Aerospace Medicine*, February 1973, pp. 184–189.

Collins, William E., James O. Boone, and Allan D. VanDeventer, "The Selection of Air Traffic Control Specialists: History and Review of Contributions by the Civil Aeromedical Institute, 1960–80," *Aviation, Space, and Environmental Medicine*, Vol. 52, No. 4, April 1981, pp. 217–240.

Collins, William E., David J. Schroeder, and Lendell G. Nye, "Relationships of Anxiety Scores to Screening and Training Status of Air Traffic Controllers," *Aviation, Space, and Environmental Medicine*, Vol. 62, No. 3, March 1991, pp. 236–240.

Colquitt, Jason A., "On the Dimensionality of Organizational Justice: A Construct Validation of a Measure," *Journal of Applied Psychology*, Vol. 86, No. 3, June 2001, pp. 386–400.

Colquitt, Jason A., Jerald Greenberg, and Cindy P. Zapata-Phelan, "What Is Organizational Justice? A Historical Overview," in Jerald Greenberg and Jason Colquitt, eds., *Handbook of Organizational Justice*, Mahwah, N.J.: Lawrence Erlbaum Associates, 2005, pp. 3–58.

Colquitt, Jason A., and John C. Shaw, "How Should Organizational Justice Be Measured?" in Jerald Greenberg and Jason Colquitt, eds., *Handbook of Organizational Justice*, Mahwah, N.J.: Lawrence Erlbaum Associates, 2005, pp. 113–154.

Commander, Air Education and Training Command, *Technical and Basic Military Training Administration*, Air Education and Training Command Instruction 36-2215, September 9, 2010. As of June 1, 2011: http://www.e-publishing.af.mil/shared/media/epubs/AETCI36-2215.pdf

Cortina, Jose M., Mary L. Doherty, Neal Schmitt, Gary Kaufman, and Richard G. Smith, "The 'Big Five' Personality Factors in the IPI and MMPI: Predictors of Police Performance," *Personnel Psychology*, Vol. 45, No. 1, March 1992, pp. 119–140.

Cramer, Phebe, *The Development of Defense Mechanisms: Theory, Research, and Assessment*, New York: Springer-Verlag, 1991.

Crooks, Terence J., "The Impact of Classroom Evaluation Practices on Students," *Review of Educational Research*, Vol. 58, No. 4, Winter 1988, pp. 438–481.

Defense Language Institute Foreign Language Center, "DLPT Frequently Asked Questions," undated. As of August 18, 2010:
http://www.dliflc.edu/dlptfaq.html

DLIFLC—*See* Defense Language Institute Foreign Language Center.

Dorynei, Zoltan, *Motivational Strategies in the Language Classroom*, Cambridge, UK: Cambridge University Press, 2001.

Driskell, James E., and Eduardo Salas, *Stress and Human Performance*, Mahwah, N.J.: Lawrence Erlbaum Associates, 1996, pp. 49–126.

Eble, Kenneth Eugene, *The Craft of Teaching: A Guide to Mastering the Professor's Art*, 2nd ed., San Francisco, Calif.: Jossey-Bass, 1988.

Ellis, Thomas I., "Extending the School Year and Day," *ERIC Clearinghouse on Educational Management: Eric Digest*, No. 7, ED259450, 1984. As of April 26, 2011:
http://www.ericdigests.org/pre-922/year.htm

Emerson, Ralph Waldo, *Self-Reliance, and Other Essays*, New York: Dover Publications, 1841 (1993).

Falomir-Pichastor, Juan M., Christian Staerklé, Marie-Aude Depuiset, and Fabrizio Butera, "Perceived Legitimacy of Collective Punishment as a Function of Democratic Versus Non-Democratic Group Structure," *Group Processes and Intergroup Relations*, Vol. 10, No. 4, October 2007, pp. 565–579.

Fashola, Olatokunbo S., *Review of Extended-Day and After-School Programs and Their Effectiveness*, Baltimore, Md.: Center for Research on the Education of Students Placed at Risk, Johns Hopkins University and Howard University, Report 24, October 1998.

Fink, Sharon Birkman, "Recruiting Across the Generation Gap," *RecruitingTrends.com*, November 21, 2007.

Finn, R. H., and Sang M. Lee, "Salary Equity: Its Determination, Analysis, and Correlates," *Journal of Applied Psychology*, Vol. 56, No. 4, August 1972, pp. 283–292.

Fisher, Julie M., "Extended School Days Considered Nationally," *Etownian*, April 5, 2007.

Form 125A—*See* Commander, Air Education and Training Command (2010).

Frand, Jason L., "The Information-Age Mindset: Changes in Students and Implications for Higher Education," *EDUCAUSE Review*, Vol. 35, No. 5, September–October 2000, pp. 14–24. As of April 26, 2011:
http://net.educause.edu/ir/library/pdf/ERM0051.pdf

Gardley, Marlon K., Ryan F. Caulk, Leonardo Cardenas, and Peter Joffe, *Virtual Environment and Technology Survey at Keesler Air Force Base Survey Report*, Randolph Air Force Base, Texas: Air Education and Training Command, Studies and Analysis Squadron, August 29, 2008.

Gardner, Susan, and Susanna Eng, "What Students Want: Generation Y and the Changing Function of the Academic Library," *Libraries and the Academy*, Vol. 5, No. 3, July 2005, pp. 405–420.

Gerber, Bud, and Mike Wilson, "Millennial Generation Traits and Teaching," briefing given at the University of North Carolina Teaching and Learning with Technology Conference, undated. As of April 26, 2011: http://conference.unctlt.org/proposals/presentations/conf4/838_Millennial%20Generation%20Tra.ppt

Greenberg, Jerald, "Determinants of Perceived Fairness of Performance Evaluations," *Journal of Applied Psychology*, Vol. 71, No. 2, May 1986, pp. 340–342.

———, "Cultivating an Image of Justice: Looking Fair on the Job," *Academy of Management Executive*, Vol. 2, No. 2, May 1988, pp. 155–158.

———, "The Organizational Waiting Game: Delay as a Status-Asserting or Status-Neutralizing Tactic," *Basic and Applied Social Psychology*, Vol. 10, No. 1, 1989, pp. 13–26.

———, "Looking Fair vs. Being Fair: Managing Impressions of Organizational Justice," in B. M. Staw and L. L. Cummings, eds., *Research in Organizational Behavior*, Vol. 12, Greenwich, Conn.: JAI Press, 1990a, pp. 111–157.

———, "Employee Theft as a Reaction to Underpayment Inequity: The Hidden Cost of Pay Cuts," *Journal of Applied Psychology*, Vol. 75, No. 5, October 1990b, pp. 561–568.

———, "The Social Side of Fairness: Interpersonal and Informational Classes of Organizational Justice," in Russell Cropanzano, ed., *Justice in the Workplace*, Vol. 1: *Approaching Fairness in Human Resource Management*, Mahwah, N.J.: Erlbaum, 1993, pp. 79–103.

Greenberg, Jerald, and Robert J. Bies, "Establishing the Role of Empirical Studies of Organizational Justice in Philosophical Inquiries into Business Ethics," *Journal of Business Ethics*, Vol. 11, No. 5–6, May 1992, pp. 433–444.

Greenberg, Jerald, and Jason Colquitt, *Handbook of Organizational Justice*, Mahwah, N.J.: Lawrence Erlbaum Associates, 2005.

Greller, Martin M., Charles K. Parsons, and Debora R. D. Mitchell, "Additive Effects and Beyond: Occupational Stressors and Social Buffers in a Police Organization," in James C. Quick, Lawrence R. Murphy, and Joseph J. Hurrell, eds., *Stress and Well-Being at Work: Assessments and Interventions for Occupational Mental Health*, Washington, D.C.: American Psychological Association, 1992, pp. 33–47.

Hardy, Lew, "A Test of Catastrophe Models of Anxiety and Sports Performance Against Multidimensional Anxiety Theory Models Using the Method of Dynamic Differences," *Anxiety, Stress and Coping: An International Journal*, Vol. 9, No. 1, 1996, pp. 69–86.

Hartson, Louis D., "A Preliminary Study of a Test for Air Traffic Controllers," *Journal of Applied Psychology*, Vol. 50, No. 2, April 1966, pp. 138–142.

Headquarters 2nd Air Force, *Airman Leader Guide*, Keesler, Miss., June 2002.

Headquarters Air Education and Training Command, *Technical and Basic Military Training Administration*, April 22, 2003.

———, *Administration of Military Standards and Discipline Training*, Instruction 36-2216, June 26, 2004.

———, *Faculty Development and Master Instructor Programs*, Instruction 36-2202, June 21, 2005.

———, *TTMS Student Management*, Randolph Air Force Base, Texas, March 2007.

Headquarters Air Force, *Trained Personnel Requirements*, Air Force Instruction 36-2616, December 10, 1993.

———, *AFSC 1C1X1: Air Traffic Control Operations*, Career Field Education and Training Plan 1C1X1 Parts I and II, January 31, 2002.

———, *AFSC 2A6X2: Aerospace Ground Equipment*, Career Field Education and Training Plan 2A6X2 Parts I and II, November 2005a.

———, *AFSC 1T2X1: Pararescue Specialty*, Career Field Education and Training Plan 1T2X1 Parts I and II, November 15, 2005b.

———, *AFSC 1C2X1: Combat Control*, Career Field Education and Training Plan 1C2X1 Parts I and II, November 1, 2006.

———, *AFSC 3E8X1: Explosive Ordnance Disposal*, Career Field Education and Training Plan 3E8X1 Parts I and II, August 2007.

Headquarters Air Force Intelligence, Surveillance and Reconnaissance Directorate, *Airborne Cryptologic Language Analyst Specialty AFSC 1A8X1 Career Field Education and Training Plan*, Washington, D.C., February 1, 2009.

Headquarters Air Force Personnel Directorate, *Air Force 5-Year Personnel Research Plan*, undated.

Headquarters Air Force Recruiting Service, "Air Force Special Tactics: It's Air Force with an Attitude—Jump In," recruiting brochure, Randolph Air Force Base, Texas, EA 02-119, undated (a).

———, "U.S. Air Force Combat Control: First There for Challenge and Adventure," recruiting brochure, Randolph Air Force Base, Texas, EA 01-025, undated (b).

Hogan, Joyce, and Robert Hogan, "Noncognitive Predictors of Performance During Explosive Ordnance Disposal Training," *Military Psychology*, Vol. 1, No. 3, 1989, pp. 117–133.

Hogan, Robert, *Hogan Personality Inventory: Manual*, Minneapolis: National Computer Systems, 1986.

Holtzman, Diane, Michael Ciocco, and Debra Dagavarian, "Teaching the Millennial Generation," briefing, Middle States Commission on Higher Education, undated.

Hom, Peter W., Loriann Roberson, and Aimee D. Ellis, "Challenging Conventional Wisdom About Who Quits: Revelations from Corporate America," *Journal of Applied Psychology*, Vol. 93, No. 1, January 2008, pp. 1–34.

Hume, David A., *Reward Management: Employee Performance, Motivation and Pay*, Oxford: Blackwell, 1995.

Jones, G., and A. Swain, "Intensity and Direction as Dimensions of Competitive State Anxiety and Relationships with Competitiveness," *Perceptual and Motor Skills*, Vol. 74, No. 2, April 1992, pp. 467–472.

Jones, Steve, and Mary Madden, "The Internet Goes to College: How Students Are Living in the Future with Today's Technology," Pew Internet and American Life Project, September 15, 2002. As of April 26, 2011: http://www.pewinternet.org/Reports/2002/The-Internet-Goes-to-College.aspx

Kamdar, Dishan, Daniel J. McAllister, and Daniel B. Turban, "'All in a Day's Work': How Follower Individual Differences and Justice Perceptions Predict OCB Role Definitions and Behavior," *Journal of Applied Psychology*, Vol. 91, No. 4, July 2006, pp. 841–855.

Keinan, Giora, Nehemia Friedland, and Vardi Sarig-Naor, "Training for Task Performance Under Stress: The Effectiveness of Phased Training Methods," *Journal of Applied Social Psychology*, Vol. 20, No. 18, October 1990, pp. 1514–1529.

Kirmeyer, Sandra L., and Thomas W. Dougherty, "Work Load, Tension, and Coping: Moderating Effects of Supervisor Support," *Personnel Psychology*, Vol. 41, No. 1, March 1988, pp. 125–139.

Knapik, J. J., S. H. Bullock, S. Canada, E. Toney, J. D. Wells, E. Hoedebecke, and B. H. Jones, "Influence of an Injury Reduction Program on Injury and Fitness Outcomes Among Soldiers," *Injury Prevention*, Vol. 10, 2004, pp. 37–42.

Kouzes, James M., and Barry Z. Posner, *A Leader's Legacy*, San Francisco, Calif.: Jossey-Bass, 2006.

Kunstler, Barton, "The Millennial University, Then and Now: From Late Medieval Origins to Radical Transformation," *On the Horizon*, Vol. 14, No. 2, 2006, pp. 62–69.

Lazarus, Richard S., and Susan Folkman, *Stress, Appraisal, and Coping*, New York: Springer Publishing Company, 1984.

Lee, Ji Hyun, "Managing the Millennial Generation," *Small Business Review*, undated. As of April 26, 2011: http://smallbusinessreview.com/human_resources/managing-millennial-generation/index.html

Lett, John, director, Research and Analysis Division, Defense Language Institute Foreign Language Center, and John Dege, director, institutional research, Defense Language Institute Foreign Language Center, "Past Attrition Statistics," briefing, in Bree E. Whelan, ed., *Workshop on Language Student Attrition*, U.S. Army Research Institute for the Behavioral and Social Sciences, Study Report 2002-02, December 2001, pp. 7–19. As of August 18, 2010:
http://www.dtic.mil/cgi-bin/GetTRDoc?Location=U2&doc=GetTRDoc.pdf&AD=ADA399533

Levenson, Hanna, Mary L. Hirschfeld, Alexander Hirschfeld, and Barbara Dzubay, "Recent Life Events and Accidents: The Role of Sex Differences," *Journal of Human Stress*, Vol. 9, No. 1, March 1983, pp. 4–11.

Leventhal, Gerale S., Jurgis Karuza Jr., and William Rick Fry, "Beyond Fairness: A Theory of Allocation Preferences," in Gerold Mikula, ed., *Justice and Social Interaction: Experimental and Theoretical Contributions, from Psychological Research*, New York: Springer-Verlag, 1980, pp. 167–218.

Liao, H., and D. E. Rupp, "The Impact of Justice Climate and Justice Orientation on Work Outcomes: A Cross-Level Multifocal Framework," *Journal of Applied Psychology*, Vol. 90, 2005, pp. 242–256.

Lind, E. Allan, "Social Justice," in Alan E. Kazdin, ed., *Encyclopedia of Psychology*, Vol. 7, 2000, pp. 346–347.

———, "Fairness Heuristic Theory: Justice Judgments as Pivotal Cognitions in Organizational Relationships," in Jerald Greenberg and Russell Cropanzano, eds., *Advances in Organizational Justice*, Stanford, Calif.: Stanford University Press, 2001, pp. 56–88.

Lind, E. Allan, and Tom R. Tyler, *The Social Psychology of Procedural Justice*, New York: Plenum Press, 1988.

Lind, E. Allan, and Kees van den Bos, "When Fairness Works: Toward a General Theory of Uncertainty Management," *Research in Organizational Behavior*, Vol. 24, 2002, pp. 181–223.

Maslach, Christina, and Ayala Pines, "The Burn-Out Syndrome in the Day Care Setting," *Child Care Quarterly*, Vol. 6, No. 2, Summer 1977, pp. 100–113.

Masten, R. L., "Leadership, Quality and the U.S. Navy," in Sarita Chawla and John Renesch, eds., *Learning Organizations: Developing Cultures for Tomorrow's Workplace*, Portland, Ore.: Productivity Press, 1995, pp. 123–145.

McCrae, Robert R., Paul T. Costa, Margarida Pedroso de Lima, António Simões, Fritz Ostendorf, Alois Angleitner, Iris Marusic, Denis Bratko, Gian Vittorio Caprara, Claudio Barbaranelli, and Joon-Ho Chae, "Age Differences in Personality Across the Adult Life Span: Parallels in Five Cultures," *Developmental Psychology*, Vol. 35, No. 2, 1999, pp. 466–477.

McMahon, Mark, and Romana Pospisil, "Laptops for a Digital Lifestyle: Millennial Students and Wireless Mobile Technologies," *Proceedings Ascilite Brisbane 2005*, 2005. As of April 26, 2011:
http://www.ascilite.org.au/conferences/brisbane05/blogs/proceedings/49_McMahon%20&%20Pospisil.pdf

Nagesh, Gautham, "Generation Y Seeks More Interaction with Government Online," *Government Executive*, July 2, 2008. As of April 26, 2011:
http://www.govexec.com/dailyfed/0708/070208n1.htm

NAS Recruitment Communications, "Generation Y: The Millennials: Ready or Not, Here They Come," Cleveland, Ohio, 2006.

Navy School Explosive Ordnance Disposal, *Training Course Control Document for Explosive Ordnance Disposal Basic (Common Core)*, Eglin Air Force Base, Fla., September 2007.

Nye, Lendell G., and William E. Collins, "Some Personality and Aptitude Characteristics of Air Traffic Control Specialist Trainees," *Aviation, Space, and Environmental Medicine*, Vol. 64, August 1993, pp. 711–716.

Oaks, David W., Gerald R. Ferris, Joseph J. Martocchio, M. Ronald Buckley, and Dana Broach, "Cognitive Ability and Personality Predictors of Training Program Skill Acquisition and Job Performance," *Journal of Business and Psychology*, Vol. 15, No. 4, Summer 2001, pp. 523–548.

Oblinger, Diana G., and Brian L. Hawkins, "The Myth About Students," *EDUCAUSE Review*, Vol. 40, No. 5, September–October 2005, pp. 12–13. As of April 26, 2011:
http://www.educause.edu/EDUCAUSE+Review/EDUCAUSEReviewMagazineVolume40/TheMythaboutStudents/158010

Oblinger, Diana G., and James L. Oblinger, eds., *Educating the Net Generation*, Boulder, Colo.: EDUCAUSE, 2005. As of April 26, 2011:
http://bibpurl.oclc.org/web/9463

Pardini, Priscilla, "Extended School Days," *School Administrator*, August 2001. As of April 26, 2011:
http://www.aasa.org/SchoolAdministratorArticle.aspx?id=10774

Pashler, Harold E., Patrice M. Bain, Brian A. Bottge, Arthur Graesser, Kenneth Koedinger, Mark McDaniel, and Janet Metcalfe, *Organizing Instruction and Study to Improve Student Learning: A Practice Guide*, Institute of Education Sciences, National Center for Education Research, U.S. Department of Education, NCER 2007-2004, September 2007. As of April 26, 2011:
http://purl.access.gpo.gov/GPO/LPS117059

Prensky, Marc, "Digital Natives, Digital Immigrants Part 1," *On the Horizon*, Vol. 9, No. 5, October 2001a, pp. 1–6.

———, "Digital Natives, Digital Immigrants, Part 2: Do They Really Think Differently?" *On the Horizon*, Vol. 9, No. 6, December 2001b, pp. 1–6.

———, "EDUCAUSE Learning Initiative," briefing, January 30, 2006. As of April 26, 2011:
http://net.educause.edu/ir/library/pdf/ELI0606.pdf

Priddy, Lynn, "The View Across: Patterns of Success in Assessing and Improving Student Learning," *On the Horizon*, Vol. 15, No. 2, 2007, pp. 58–79.

Principals' Partnership, "Research Brief: Extended School Day," undated.

Raines, Claire, "Managing Millennials," in Claire Raines, *Connecting Generations: The Sourcebook for a New Workplace*, 2002.

Ramos, Robert A., Michael C. Heil, and Carol A. Manning, *Documentation of Validity for the AT-SAT Computerized Test Battery: Final Report*, Washington, D.C.: U.S. Department of Transportation, Federal Aviation Administration, Office of Aviation Medicine, DOT/FAA/AM-01/6, March 2001. As of April 26, 2011:
http://purl.access.gpo.gov/GPO/LPS13995

Ricigliano, L., "After X Comes Y: Teaching the Next Generation," paper presented at the LOEX annual conference, Houston, Texas, March 13, 1999.

Sandfort, Melissa H., and Jennifer G. Haworth, "Whassup? A Glimpse into the Attitudes and Beliefs of the Millennial Generation," *Journal of College and Character*, Vol. 3, No. 3, 2002.

Sandusky, Sue Ann, "Five Lines of Effort: Commandant's Address," *Dialog on Language Instruction*, Vol. 20, No. 1–2, 2009.

Schaubroeck, John, Douglas R. May, and William F. Brown, "Procedural Justice Explanations and Employee Reactions to Economic Hardship: A Field Experiment," *Journal of Applied Psychology*, Vol. 79, No. 3, June 1994, pp. 455–460.

Schmidt, Norman B., Darin R. Lerew, and Robert J. Jackson, "The Role of Anxiety Sensitivity in the Pathogenesis of Panic: Prospective Evaluation of Spontaneous Panic Attacks During Acute Stress," *Journal of Abnormal Psychology*, Vol. 106, No. 3, August 1997, pp. 355–364.

Schmidt, Steven W., "The Relationship Between Satisfaction with Workplace Training and Overall Job Satisfaction," *Human Resource Development Quarterly*, Vol. 18, No. 4, Winter 2007, pp. 481–498.

Secretary of the Air Force, *Enlisted Classification*, Air Force Manual 36-2108, October, 2004.

Sheppard, Blair H., Roy J. Lewicki, and John W. Minton, *Organizational Justice: The Search for Fairness in the Workplace*, New York: Lexington Books, 1992.

Simons, Tony, and Quinetta Roberson, "Why Managers Should Care About Fairness: The Effects of Aggregate Justice Perceptions on Organizational Outcomes," *Journal of Applied Psychology*, Vol. 88, No. 3, June 2003, pp. 432–443.

Staal, Mark A., Amy E. Bolton, Rita A. Yaroush, and Lyle E. Bourne Jr., "Cognitive Performance and Resilience to Stress," in Brian J. Lukey and Victoria Tepe, eds., *Biobehavioral Resilience to Stress*, Boca Raton, Fla.: CRC Press, 2008.

Sweeney, Richard T., "Reinventing Library Buildings and Services for the Millennial Generation," *Library Administration and Management*, Vol. 19, No. 4, Fall 2005, pp. 165–175.

Tilley, A., and S. Brown, "Sleep Deprivation," in Dylan M. Jones and A. P. Smith, eds., *Handbook of Human Performance*, Vol. 3: *State and Trait*, London: Academic Press, 1992, pp. 237–238.

Tomlinson, Edward C., and Jerald Greenberg, "Discouraging Employee Theft by Managing Social Norms and Promoting Organizational Justice," in Roland E. Kidwell Jr. and Christopher L. Martin, eds., *Managing Organizational Deviance*, Thousand Oaks, Calif.: Sage Publications, 2005, pp. 211–236.

Trites, David K., "Problems in Air Traffic Management VI: Interaction of Training-Entry Age with Intellectual and Personality Characteristics of Air Traffic Control Specialists," *Aerospace Medicine*, December 1964, pp. 1184–1194.

Trites, David K., and Bart B. Cobb, "Problems in Air Traffic Management," *Aerospace Medicine*, May 1964, pp. 428–436.

Trites, David K., Adolph Kurek, and Bart B. Cobb, "Personality and Achievement of Air Traffic Controllers," *Aerospace Medicine*, Vol. 38, No. 11, November 1967, pp. 1145–1150.

U.S. Air Force Directorate of Personnel, "Air Force 5-Year Personnel Research Plan," Washington, D.C., undated.

U.S. Department of Defense, Uniform Code of Military Justice, Subchapter III, Nonjudicial punishment, Section 815, Article 15, Commanding officer's nonjudicial punishment. As of April 27, 2011: http://www.au.af.mil/au/awc/awcgate/ucmj.htm

U.S. General Accounting Office, *DoD Service Academies: More Changes Needed to Eliminate Hazing*, Washington, D.C., GAO/NSIAD-93-36, 1992. As of April 26, 2011: http://archive.gao.gov/d36t11/148057.pdf

———, *Military Attrition: Better Data, Coupled with Policy Changes, Could Help the Services Reduce Early Separations—Report to the Chairman and Ranking Minority Member, Subcommittee on Personnel, Committee on Armed Services, U.S. Senate*, Washington, D.C., GAO/NSIAD-98-213, September 1998. As of April 26, 2011: http://purl.access.gpo.gov/GPO/LPS16533

Velada, Raquel, António Caetano, John W. Michel, Brian D. Lyons, and Michael J. Kavanagh, "The Effects of Training Design, Individual Characteristics and Work Environment on Transfer of Training," *International Journal of Training and Development*, Vol. 11, No. 4, December 2007, pp. 282–294.

Walker, Shaundra, "Academic Library Services for the Millennial Generation," *Georgia Library Quarterly*, Vol. 43, No. 2, Summer 2006, pp. 8–12.

Weissmuller, Johnny J., "'Personality' and Mission Effectiveness," presentation at the Personality and the Military Symposium II 48th International Military Testing Association Conference, Kingston, Ont., October 4, 2006.

Whelan, Debra Lau, "SLJ Talks to Researcher Elena Silva About Extending the School Day," *School Library Journal*, February 14, 2007. As of April 26, 2011: http://www.schoollibraryjournal.com/article/CA6416222.html

Wyatt, Thomas C., and Reuven Gal, *Legitimacy and Commitment in the Military*, New York: Greenwood Press, 1990.